Plane Crash

Plane Crash

The Forensics of Aviation Disasters

George Bibel and
Captain Robert Hedges

Johns Hopkins University Press

Baltimore

Johns Hopkins University Press
2715 North Charles Street
Baltimore, Maryland 21218-4363
www.press.jhu.edu

Library of Congress Cataloging-in-Publication Data

Names: Bibel, G. D. (George D.), author. | Hedges, Robert, 1964– author.
Title: Plane crash : the forensics of aviation disasters / George Bibel and
    Captain Robert Hedges.
Description: Baltimore : Johns Hopkins University Press, [2018] | Includes
    bibliographical references and index.
Identifiers: LCCN 2017021044| ISBN 9781421424484 (hardcover : alk. paper) |
    ISBN 9781421424491 (electronic) | ISBN 1421424487 (hardcover : alk. paper)
    | ISBN 1421424495 (electronic)
Subjects: LCSH: Aircraft accidents—Investigation. | Aircraft accidents—Case
    studies.
Classification: LCC TL553.5 .B5222 2018 | DDC 363.12/465—dc23
  LC record available at https://lccn.loc.gov/2017021044

A catalog record for this book is available from the British Library.

*Special discounts are available for bulk purchases of this book. For more information, please
contact Special Sales at 410-516-6936 or specialsales@press.jhu.edu.*

Johns Hopkins University Press uses environmentally friendly book materials,
including recycled text paper that is composed of at least 30 percent post-consumer
waste, whenever possible.

*To the two greatest teachers with whom I have studied:
my father, Dr. James S. Hedges, who taught me the love of
reading, and Dr. John E. Kiser, who taught me the love
of writing*

Bob Hedges

Contents

Preface

In spite of writing a book about aviation mishaps, we want to stress that commercial aviation is by far the safest mode of transportation, and it is becoming safer all the time, in part because of comprehensive accident analysis. Today, commercial flight is so safe that airlines practically have to invent a new sequence of events to crash a plane. Most of the accident scenarios in this book are exceedingly rare if not downright unique. A thorough investigation and implementation of remedial actions significantly reduce the probability of an already remote event repeating.

This is a science book that tells stories, or a storybook that teaches science. The airplane accidents discussed here have been selected both because they touch on well-documented (and understandable) science and because they are compelling and interesting stories. Surprisingly, most accidents are survivable because planes typically crash from lower altitudes and speeds associated with a less-than-perfect takeoff or landing.

This book is a companion to my previous book, *Beyond the Black Box: The Forensics of Airplane Crashes*. It describes all new accidents and includes important contributions from an experienced airline pilot.

We are both interested in how things work on the airplane, or how things do not work during accident scenarios. Airliner performance is also of interest. Information is highly proprietary, yet data leak out in the accident reports. The book is organized around the phases of flight: takeoff, climb, cruise, approach, and landing. A disproportionate number of events from the cruise phase are presented, not because they are most common but because they are most interesting.

Introducing Captain Hedges

Robert Hedges is a Captain, line check airman, and instructor at a major US airline. Bob took his first flying lesson at the age of 14 and flew light aircraft throughout high school and college. He graduated from the University of North Carolina–Charlotte in 1986 with a degree in biology. He attended US Air Force Undergraduate Pilot Training at Williams Air Force Base, Arizona, flying the T-37 and T-38, and he subsequently flew and instructed in the C-9A (a military DC-9 used for aeromedical transportation) at Scott Air Force Base, Illinois. After leaving the USAF, he started his airline career as a B-727 Flight Engineer (Second Officer) in 1991. In the 25 years since then, he has flown and instructed in numerous aircraft, including the Airbus A320, McDonnell Douglas DC-9 and MD-80 series, Boeing B-727, B-737, B-747-400, B-757, and B-767, and the Lockheed L-1011. He has amassed over 15,000 hours of flying time and another 3,800 hours instructing in the simulator.

Bob has always had a keen interest in safety and has taught classes on human factors, crew resource management, and threat and error management. He has a master of aeronautical science degree with distinction and a graduate certificate in aviation/aerospace safety systems from Embry-Riddle Aeronautical University. Bob has taught a variety of aerospace courses at the graduate level, and he has instructed pilots from more than 30 airlines worldwide. In addition to flying and teaching in the classroom, simulator, and aircraft, he has also been a check airman, a line check airman, and a designated pilot examiner for the Federal Aviation Administration (FAA).[1]

Bob holds the following FAA certificates and ratings: Airline Transport Pilot, Airplane Single and Multiengine Land, A320, B-727, B-747-4[00], B-757, B-767, DC-9, and L-1011. He also holds Certified Flight Instructor and Flight Engineer-Turbojet certificates. He has been the production manager of a chemical manufacturing plant; studies industrial safety in the aviation, maritime, chemical processing, and nuclear power industries; and is a member of the American Nuclear Society. He enjoys researching and writing about aerospace history and safety, with particular interest in the F-105, L-1011, and the Apollo programs.

Although the book is focused on science, Captain Hedges, being a pilot, is more attuned to what is commonly called "human factors"—factors related to human error. With invaluable review and comments, Captain Hedges' contributions to this book greatly exceeded the words he wrote.

George Bibel

Plane Crash

1

Takeoff!

With just one plane, MK Airlines began operations in 1990. Their mission was to fly fresh pineapples from Africa to Europe. With business growing 30% annually, the company was adding nearly one plane each year. By 2004, MK Airlines had expanded to twelve planes—six Boeing 747s and six DC-8s. Retired passenger planes are often repurposed as cargo planes. The Douglas DC-8, introduced in 1959, was last manufactured in 1972.

On October 14, 2004, MK Airlines Flight 62, a Boeing 747-200 manufactured in 1980, took off from Halifax, Nova Scotia, shortly before 4:00 a.m. The plane crashed just outside the airport security perimeter. Air traffic controllers saw a fireball and activated the crash alarm. Fire trucks arrived on the scene in about five minutes, and 60 firefighters spent five hours putting out the main fire. The entire crew of seven was killed.

The plane clipped treetops and utility poles and cut off electric power to the airport. Emergency generators automatically restored power within a few seconds. The control tower, with its own power supply (and backup system), didn't miss a beat.

First reports of an explosion occurring before the crash did not pan out. Nevertheless, Canadian police took charge of the initial investigation. Eyewitness accounts reported the next day: "The aircraft basically didn't take off. She continued her [takeoff run] and ran off the runway and ran into the woods."[1] "All I saw was the nose going up and it looked like it was dragging and then the [airport] power went out and then you just saw white and orange sky."[2] Another eyewitness reported two explosions. "That was a quick one followed by a second one that was bigger. And then we saw a very bright orange light and I mean bright. It took up the whole sky."[3]

The next day, an airport worker said the plane entered the runway more than 2,000 feet from the end. "From where she left, they only had about 6,000 feet and it just wasn't enough runway."[4] Security videos did not support this version of events.

It was soon established that the tail struck the runway twice and dragged, and then broke off after striking an earthen berm located 1,150 feet past the end of the runway. Mounted on a concrete slab on top of the berm were the antennas for the instrument landing system. The berm was 11.6 feet high. Because of ground slope, the top of the berm was the same height as the end of the runway. (The earthen berm met all airport certification standards and was not considered a factor in the investigation.) Per mandated rules that define a safe takeoff, the lowest part of the plane (the extended landing gear) must be at least 35 feet off the ground by the end of the runway. Takeoff parameters are calculated to meet this 35-foot requirement.

Figure 1.1 shows the fuselage striking the berm. The fuselage joint of the aft (rear) pressure bulkhead was embedded in the earthen berm. Geometry requires the impact angle to be between 15° and 24°. If the aircraft pitch angle was less than 15°, the wheels would have struck the top of the berm; if greater than 24°, the tire wouldn't have struck the antenna. The tail broke off first. Immediately past the berm were the vertical and horizontal stabilizers and other tail debris. The remaining parts of the aircraft continued airborne for about another 1,200 feet before striking the ground and burst-

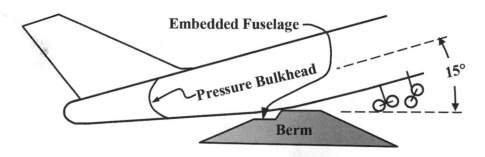

Figure 1.1. Minimum impact angle of the MK Airlines Boeing 747 as determined by the crash investigators.

ing into a fireball. The main impact was approximately 2,500 feet past the end of the runway.

The plane and crew were scheduled to fly from Luxembourg to Connecticut to Halifax to Spain and finally back to Luxembourg within 30 hours. For that reason, the three-man crew (Captain, First Officer, and Flight Engineer) had an additional Captain and Flight Engineer. Also on board was a load master and maintenance technician. The crew's remains were later identified by using DNA and dental records.

The Canadian crash investigators quickly announced that many things can cause a tail strike, including but not limited to engine failure, excess weight, and cargo shift. Two days later, it was stated that two of the four engines had been replaced in the previous month. Engine defects eventually proved to be another "red herring."

The flight data recorder was recovered on October 17, followed by the cockpit voice recorder the next day. Unfortunately, no voice recordings were recovered. The older-style magnetic tape melted in the intense fire. Newer solid-state recorders are more fire resistant.

On October 22, investigators announced that the plane had been unable to properly take off because of too little speed or too little engine thrust for the airplane's weight. On that same day, Canadian officials issued a safety advisory about procedures for weighing cargo. It was later determined the plane was about 4,400 lbs heavier than officially documented, a relatively minor error for this 747-200 and its certified maximum takeoff weight of 833,000 lbs. Despite there being no formal announcement regarding excess weight, the October 24 headlines of the London *Sunday Times* declared, "Overloading Blamed for Plane Crash That Killed Four Britons." In mid-December, the Canadian Transportation Safety Board announced that an incorrect engine throttle setting had become the focus of the investigation.

A fully loaded 747 is approximately 42% fuel (assuming the fuel tanks are full) and 42% airplane structure. That only leaves about 16% for passengers and freight, making air transportation an expensive way to haul cargo.[5]

Almost 18 months after the crash, the formal accident report was released. The flight data recorder, wreckage distribution, and runway and ground scars were used to reconstruct the plane's trajectory and accident time line. Takeoff began at 3:53:22 a.m. local time. Twenty-four seconds later (and about 1,800 feet from the beginning of the runway), the plane had accelerated to 80 knots.[6] The engines' throttles had been gradually advanced from a ground idle engine pressure ratio (or EPR, which is the ratio of engine exhaust to inlet air pressure) of 1.0 to an EPR of 1.3.

There is no direct reading of engine thrust inside the cockpit. Instead, the pilot advances the engine throttle lever for each of four engines until the EPR reaches a predetermined value calculated by the Boeing Laptop Tool for a safe takeoff.[7] EPR directly correlates with engine thrust. The Boeing software considers the takeoff weight, ambient conditions, specific runway, and the mandated requirements for a safe takeoff (see Chapter 2).

At 3:54:08 (46 seconds after the takeoff roll began), the plane was 5,500 feet down the runway and moving at 130 knots; 3,300 feet of runway remained. The takeoff procedure calls for a rotation of 12°. A plane rotates, or raises its nose, to increase lift by increasing the wing's angle of attack. During rotation, the lift force nearly triples.

The tail of Flight 62 first struck the concrete 830 feet from the end of the runway and scraped for the next 125 feet. With about 600 feet of runway remaining, and just 58 seconds after starting the takeoff roll, the pilot advanced the throttles to 92% of full throttle (corresponding to an EPR of 1.6). Advancing the throttles indicated that the pilot was aware that something was wrong with the takeoff. Because of the engine's large rotational inertia, there is a delay of perhaps 3–4 seconds between throttle advance and increased engine thrust. The tail broke off just 7 seconds later.

The plane passed the end of the runway at 152 knots and a nose-up attitude of 11.9°. The tail struck again 412 feet from the runway's end and remained in contact until 825 feet past the end of the runway. The plane became airborne just 325 feet before the tail struck the earthen berm and broke off. The second tail scrape was initially 3 inches wide, expanded to 24 inches at the end of the concrete, and dug into the dirt until gradually disappearing. There were aluminum shavings on the concrete runway along the second tail strike. The locations of the tail strikes are shown in Figure 1.2.

After ruling out a variety of scenarios, the investigators focused on the weight entered into the plane's flight software. Instead of entering the correct weight, estimated to be 780,000 lbs, the pilot accidentally entered

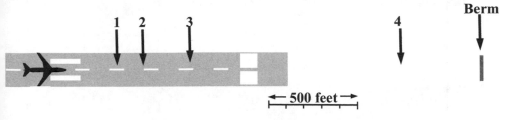

Figure 1.2. *1 and 2,* Start and end of first tail strike. *3 and 4,* Start and end of second tail strike. Drawing is approximately to scale. Runway is 200 feet wide, and wingspan is 195 feet. The paved runway is 8,800 feet long, with an additional clearway of 1,000 feet.

528,630 lbs, the takeoff weight of the previous flight. The software output instructed the pilots to set the engine pressure ratio at 1.3, the same as the previous flight. The correct EPR should have been 1.6. The plane did in fact take off and fly for 325 feet. The investigation concluded the plane would have safely flown if the tail hadn't struck (and broken off) on the earthen berm.

A heavier plane requires more engine thrust to safely take off. Or, for the same engine thrust, a heavier plane requires a longer runway. The Federal Aviation Administration–mandated runway length for the Boeing 747-200 (with the same engines as the accident plane), which meets all the requirements for a safe takeoff, is shown in Figure 1.3. Adjustments must be made for ambient conditions, runway, slope, and wind.

Normal Takeoff Procedure

Performance calculations are required for every takeoff because runway lengths vary by thousands of feet, takeoff weight changes every flight, and lift forces are sensitive to ambient conditions. The normal takeoff procedure identifies three speeds: V_1, V_R, and V_2. V_1 is called the decision speed and represents the maximum speed at which the pilot can safely abort the takeoff. Above V_1, there is not enough remaining runway to safely stop. Once V_1 speed is reached, it is safe to continue the takeoff even if one engine abruptly fails.

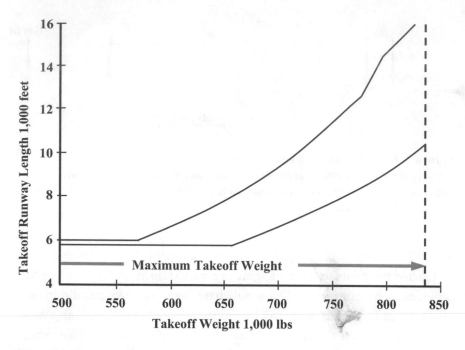

Figure 1.3. Takeoff distance mandated by the Federal Aviation Administration for a 747-200 with JT9D-7Q engines at 59°F, zero wind, and no runway gradient. Shown are two runways, one at sea level (lower curve) and one at 6,000 feet (upper curve).

At V_R, or rotation speed, the pilot pulls back on the control column in a continuous motion to raise the nose. V_2 is officially known as the takeoff safety speed. The correct V_2 speed also ensures an acceptable rate of climb and that other safety standards are met for aircraft handling and controllability—even with one engine out. The takeoff speeds are precalculated and presented for various scenarios in the flight manual or are alternately calculated by software. The three speeds are fundamental to safe flight and become part of the flight plan. Additional takeoff requirements are given in Chapter 2.

Airliners have two pilots: a Captain and First Officer (FO, formerly called a copilot). The pilots normally take turns flying the plane on alternating flights. The pilot at the controls is called the Pilot Flying (PF); the other pilot is called the Pilot Monitoring (PM). One of the PM's biggest tasks is to monitor the status of the aircraft and to ensure that the PF is flying the aircraft correctly. During takeoff, the PF flies the plane. The PM monitors the speeds and announces V_1 and V_R as they occur. The takeoff sequence is a five-step process: (1) accelerate to V_1, (2) PM announces V_1, (3) accelerate to V_R, (4) PM announces V_R, and (5) rotation at V_R. If everything goes as planned, the plane lifts off on its own close to the predicted V_2 speed.

The correct takeoff speeds for the Halifax accident flight were $V_1 = 150$ knots, $V_R = 162$ knots, and $V_2 = 172$ knots. The incorrect speeds, used for a plane that weighed 250,000 lbs less than the actual takeoff weight of 780,000 lbs, were $V_1 = 128$ knots, $V_R = 128$ knots, and $V_2 = 137$ knots.

Lift

Fundamentally, the wings must provide sufficient lift to get the plane off the ground. One explanation of lift is the downward deflection of air. This downward deflection can be seen by blowing smoke over a stationary wing in a wind tunnel. If wind in a tunnel is blown at 300 mph across a wing and deflected downward air is blown at 10°, the component of downward deflected air is 52 mph, as shown in Figure 1.4. Why does the airflow deflect downward and follow the wing's contour? If the air flow on the top surface was straight back horizontally, without downward deflection, the air molecules would drag along molecules underneath. Those molecules dragged along would in turn drag additional air molecules underneath and so on. The zone immediately above the wing, now depleted of air molecules, experiences reduced pressure. The higher pressure above the wing deflects the airflow downward along the wing's surface.

It takes a force to deflect the air downward. The equal and opposite force appears as lift on the wings. If 100 lbs of air per second flows off the wing with a downward component of 52 mph, the lift can be calculated with Newton's second law ($f = ma$) to be 237.5 lbs of force. Using the same logic, a propeller that accelerates 100 lbs of air per second to a speed of 76 ft/sec will create 237.5 lbs of thrust. Describing lift as the downward acceleration of air explains a few simple concepts about wings.

If the plane's speed doubles, the amount of air deflected doubles and the downward speed also doubles; hence lift increases with airspeed squared. More lift is also created if the wing increases its angle of attack (AOA) by pitching the plane's nose up. Lift scales with AOA; that is, if the AOA doubles, the downward velocity of the air doubles, and so does the lift.[8] And if the AOA triples, so does lift. If the AOA is increased beyond a critical limit, the "flow separates" (official engineering jargon) and doesn't follow the wing's contour. The flow is no longer deflected downward, creating lift; instead, it flows off the wing in a turbulent wake. The wing is said to have stalled (see Chapter 4). Lift has not been historically calculated by the amount of downward deflected air, mainly because there has been no easy way to calculate the amount of air being deflected. And not all the air is deflected in the same direction. On the wing's leading edge, the air is deflected up, over, and around.

Today, computer simulations can predict air deflections; surface pressures are instead directly calculated. But modern airplane manufacturers sign billion-dollar contracts before any prototype is built and flown. The contract guarantees performance, specifically fuel efficiency, and specifies penalties for fuel efficiency that is less than promised. A 1% error could cost $500,000 per plane per year, a nasty surprise for any airline that buys 10 new planes to save fuel. For that reason, wind tunnel testing is still used to fine-tune the computer models.

Lift has been historically described with the following equation:

$$\text{lift} = \frac{1}{2} \times (\text{coefficient of lift}) \times (\text{air density}) \times (\text{area of wings}) \times \text{airspeed}^2$$

The coefficient of lift is determined experimentally for different angles of attack. This simple equation predicts how lift changes with varying air density and airspeed.

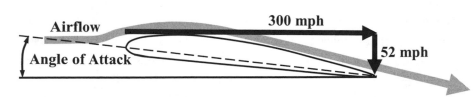

Figure 1.4. Airflow at 300 mph air deflected downward 10° creates a 52 mph vertical component of downward air velocity.

Downward Airflow and Lift

In takes a force of 237.5 lbs to accelerate 100 lbs/sec of air to a speed of 52 mph (76 ft/sec) down. Acceleration is the rate of change of velocity. The still air changes its vertical velocity from 0 ft/sec to 76 ft/sec in one second. The acceleration of the air is (0 – 76) ft/sec divided by one second, or –76 ft/sec^2. The negative sign indicates a downward acceleration. Per Newton's law, mass × acceleration equals force. The 100 lbs of air × the acceleration equals the force that the wing exerts on the air to change its direction. The air's weight must be converted into a mass per Newton's Law. (For a falling mass, Newton's law becomes weight = mass × gravitational acceleration, which defines mass to be the weight divided by the acceleration of gravity.)

The complete calculation is:

$$100 \text{ lbs} \times \frac{\sec^2}{32 \text{ feet}} \times \frac{-76 \text{ feet}}{\sec^2} = -237.5 \text{ lbs of force down.}$$

The equal and opposite positive force (the force the air exerts on the wings) appears as lift on the wings.

The density of air is sensitive to changes in altitude and temperature. The density of "standard air" (59°F at sea level) is 0.076474 lbs/foot^3, about 2 lbs per cubic yard. Two extremes might be Anchorage (elevation 151 feet) at –40°F (0.094 lbs/foot^3) and Denver (elevation 5,431) at 100°F (0.0577 lbs/foot^3)—a swing of ±24%.

By international agreement, air pressure at sea level is defined to be 14.6959 psi at 59°F. This corresponds to a barometric reading of 29.92 inches of mercury. (An airplane's altimeter is in fact a device that measure air pressure.) Actual air pressure varies with movement of weather systems. For example, a pressure of 28.00 inches of mercury is equivalent to an altitude of about 1,800 feet above sea level. To account for changes in air density, pilots enter into the flight computer (or use charts in the flight manual to adjust the takeoff parameters) the barometric pressure and air temperature. Air pressure also affects engine performance. Less air density also corresponds to less air entering the engines (jet or piston) and a reduction of power.

If the plane is rolling down a level runway (wheels still on the ground), the angle of attack (AOA) on a Boeing 747 is 2° (i.e., the wing's angle of incidence is said to be 2°). To increase the AOA during climb out, the plane's nose is pitched up relative to the direction of flight. The plane's "pitch" or "pitch attitude" (also referred to as just "attitude") is the angle of the plane's longitudinal axis with the horizontal. The wing's AOA is relative to the direction of flight, as shown in Figure 1.5. The AOA equals the pitch angle minus the direction of flight plus the angle of incidence. If the same plane shown in Figure 1.5 maintains the same direction of flight, but pitches the nose up 5°, the angle of attack increases to from 15° to 20°.

The coefficient of lift (for standard air) for the 747-100, the earliest version of the 747 (ca. 1970), which is geometrically similar to the 747-200 Halifax accident plane, is shown in Figure 1.6. Figure 1.6 shows the lift coefficient for flaps up and flaps at 20°.

The wings are designed to gradually stall, allowing the pilot to respond to any upsets. A pitch angle greater than 25° is considered an upset condition. The pilot can easily recover from a 25° pitch; however, the normal margins for safe flight are compromised.

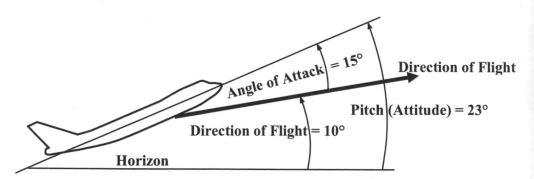

Figure 1.5. Angle of attack equals pitch angle minus the flight path angle plus the angle of incidence. The Boeing 747's 2° angle of incidence is not shown.

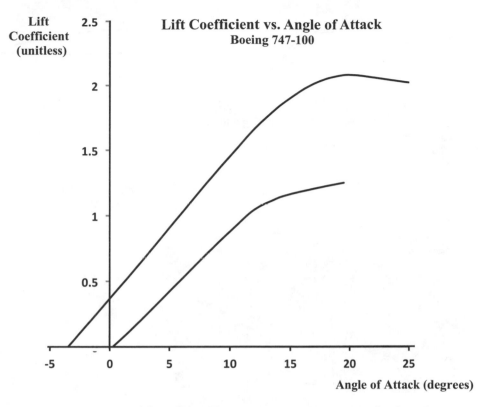

Figure 1.6. Basic lift coefficient for a Boeing 747-100 versus angle of attack. Upper curve is for Flaps 20, and the lower curve for flaps up.

Flaps increase the 747's wing area by nearly 24%. Flaps also increase the downward deflection of air. The Boeing 747 has triple-slotted flaps and leading edge flaps, as shown in Figures 1.7 and 1.8. The 747 has seven different flap settings: 1, 5, 10, 15, 20, 25, and 30. The numbers refer to greater extension and increased downward angle. The Halifax accident occurred with flaps set at 20. Planes have crashed with the improbable triple whammy of both pilots forgetting to deploy the flaps and slats on takeoff, along with failure of the "flaps not deployed" takeoff warning.[9]

Almost everything associated with flight changes the airflow over the wings. Airplane designers consider these changes by modifying the coef-

Figure 1.7. Leading edge flaps and trailing edge triple flaps shown in open and closed positions.

Figure 1.8. Triple-slotted flaps fully extended for landing for a Boeing 747. Leading edge flaps are more difficult to see. Most airliners have leading edge slats, though some like the Boeing 747 have leading edge flaps. Wikimedia/Arpingstone

ficient of lift with "adjustments." There are adjustments for wing flex, airspeed (above about 300 mph, air density changes as it flows over the wing), ground effect, deflection of the control surfaces, and many more. Nevertheless, rough estimates of lift for the Halifax accident flight can be made.

Without flaps extended, the lift coefficient for the 747 (moving on level ground) with an angle of attack of 2° is about 0.15. Using the correct takeoff speed of 172 knots or 290 ft/sec, an air density of 0.078 lbs/ft³, and a wing area of 5,500 ft², the lift equation predicts a lift force of:

$$\tfrac{1}{2} \times 0.15 \times 0.078 \, \frac{lbs}{ft^3} \times \frac{sec^2}{32.2 \, ft} \times 5,500 \, ft^2 \times \frac{290^2 \, ft^2}{sec^2} = 84,000 \, lbs$$

There is simply not enough lift to take off. Deploying flaps to 20 increases the coefficient of lift to 0.58. Lift nearly quadruples—still not enough lift for the 780,000 lb plane to take off. The plane must rotate to increase the angle of attack. With takeoff rotation expected at 10°, the angle of attack is 10° + 2° = 12°; the coefficient of lift increases to 1.65. The lift force is now estimated to be 924,000 lbs, more than enough to lift the 780,000 lb plane. The additional force is required to accelerate the plane upward, to overcome air resistance and the downward tail force that rotates the plane. During a normal takeoff, the plane's speed continues to increase and create more lift.

Because lift is proportional to speed squared, the incorrect takeoff speed, 137 knots, will provide $(137/172)^2 = 0.63$, or only 63% of the lift required for a safe takeoff. The accident plane was in fact airborne at a speed between 152 and 155 knots. The Halifax accident plane had a tail strike at about 12° nose up; this makes the maximum angle of attack (AOA) on takeoff 14°. Our crude estimate of lift at 152 knots and a tail strike AOA just barely provides enough lift to support the plane's weight. The Boeing reference increases the coefficient of lift for ground effect[10] by about 13%.

The tail scrape created an additional force that slowed the plane. Because of all the adjustments required to calculate an actual coefficient of lift, the numbers presented here are estimates. Nevertheless, they are consistent with the facts of the accident. Boeing states that the minimum takeoff speed for the accident plane (including the increased AOA from a tail strike) is 150 knots ± 2 knots.

The slowest takeoff speed occurs at the highest angle of attack. This speed, known as V_{MU}, is established with test flights by raising the nose until the tail grazes the runway. Federal Aviation Administration rules require the takeoff speed to be at least 110% of V_{MU}.

Captain Hedges Explains: Taking Off from Denver

As altitude and temperature increase, air molecules move farther apart. For this reason, an aircraft will perform (climb) worse in "hot and high" situations than on a cold day at lower elevations. This is why airports like the one in Denver have extremely long runways. At an elevation of 5,431 feet above sea level, Denver International Airport is starting with a performance disadvantage. In the summer, when temperatures can routinely exceed 90°F, aircraft performance is extremely compromised, often to the point that payload reductions need to be taken to make the departure legal. (This is obviously very undesirable, as removing cargo, luggage, passengers, or fuel is operationally problematic.) To get the aircraft going fast enough to be able to climb at V_2, the aircraft is likely to require less flap deflection (and hence less drag), which makes the takeoff distance extremely long. The airplane must accelerate on the runway in suboptimum conditions to a higher V_1 speed (the less flaps used, the higher the airspeed required to produce lift is). This is why Denver has a runway that is an astonishing 16,000 feet long. Frequently the single thing limiting takeoffs at Denver is the maximum tire limit speed. Each type of tire has a maximum speed at which it can safely turn without the risk of flying apart. (Catastrophic tire failure led to the loss of an Air France Concorde on July 25, 2000, in Paris, and the eventual shuttering of the Concorde program.) V_1 speed obviously cannot exceed that speed, so tire limit speed can become the limiting factor in places like Denver, where the ability to climb is likely to be limited.

Rotation

The main landing gear on the massive 747 has four sets of four tires under the wings. During rotation, a downward force is created in the horizontal stabilizer in the tail section, and the aft main landing gear acts as a pivot, as shown in Figure 1.9. The plane rotates with a lever action similar to a giant teeter-totter.

When the tail section suddenly broke off MK Airlines Flight 62, the 747 lost the downward tail force. Much like a person jumping off the down seat on a teeter-totter, the person in the air rapidly rotates down. In the Halifax accident, the plane rapidly rotated (after the tail broke off) from its maximum nose-up attitude of +14° to –20°. This resulted in a violent impact and fireball. A shallower impact angle might have been survivable—the pro-

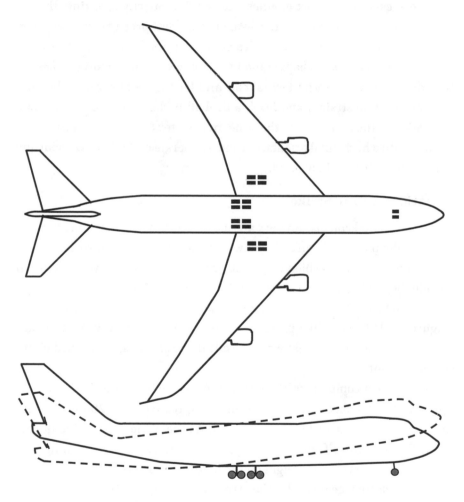

Figure 1.9. Rotation of the Boeing 747.

verbial hit and skip (shallow impact angle) instead of hit and stick (steep impact angle) crash that did occur.[11] Most airplane crashes occur during takeoff or landing and are in fact quite survivable because the impact angle is significantly less than 20°.

Fragmentation damage of the plane is often related to the vertical downward speed, or sink rate, and impact angle. Assuming the plane was traveling at 155 knots and suddenly rotated 20°, the estimated sink rate is about 89 ft/sec. A severe impact that fragments the plane and creates fuel misting often results in a fireball. Misting creates an explosive fuel/air mixture that can be sucked into the jet engines and ignited. Strictly speaking, the explosion is a subsonic deflagration—the flame front propagates at less than the speed of sound. Other examples of impact, followed by a fireball, are given at the end of this chapter and in Chapter 3. Because most crashes occur during a less-than-perfect takeoff and landing, and because the pilot is mostly flying straight and level, a fireball is the exception rather than the rule. On the other end of the scale, in Chapter 8 describes a Boeing 777 struck with a horizontal orientation and a sink rate of 25 ft/sec without any fuel misting or fire of any kind.

Rotation and Tail Strike

If the wrong takeoff parameters are used, if rotation is not done properly, or if the plane is disturbed by a wind gust, the plane can strike its tail. The plane sits on a cushion of pressurized nitrogen (and hydraulic fluid) inside the landing gear shock struts. The Boeing 747 inner shock strut cylinder extends or telescopes up to 26.5 inches within the outer cylinder (see Figure 1.10). The internal pressure in the landing gear prevents the hard metal-on-metal impact that would result if the shock strut reached either extreme limit.

As the strut compresses, the nitrogen pressure increases. Sitting on the runway, the 747 shock strut is typically compressed about 20 inches, with a nitrogen pressure of 1,600 to 1,700 psi. As the plane gains speed during takeoff, the wing's lift force increases and the weight on the landing gear decreases. With reduced weight on the landing gear, the nitrogen pressure in the shock strut begins to lift the plane even though the wheels are still on the ground.

The axles for each pair of wheels are attached to a horizontal truck beam. The truck beam, attached to the vertical shock strut with a pivot, is free to rotate. The shock strut cylinders, truck beam, pivot point, and axles are shown schematically in Figure 1.10.

Because the shock strut extends, and the truck beam rotates, the geometry of tail strike is not precisely defined. In fact, the plane's pivot point during rotation moves from the center of the truck beam to the center of the back wheel. Some of this motion can be seen in an amazing photo taken by Roel Kroes (see Figure 1.11).

Boeing identifies two extremes for tail contact on the 747-200 and a range of potential contact locations occurring along a 58-foot section of the tail. The first extreme is zero tilt (all wheels on the ground) with normal shock strut compression (i.e., the plane's weight compresses the strut about 20 inches); tail strike occurs if the nose pitches up more than 11.1°. The other extreme is where the landing gear has maximum tilt (only one wheel contacts the ground) and the shock strut is fully extended; in this case, contact occurs for rotations greater than 13.1°. Compression of the landing gear shock strut, about 26.5 inches, is yet another variable. The inner workings of the shock strut are discussed in Chapter 8.

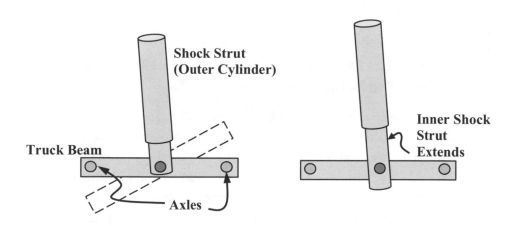

Figure 1.10. The truck beam attaches to the shock strut with a pivot and is free to rotate. The inner shock strut extends up to 26.5 inches.

Figure 1.11. Articulated main landing gear on a Boeing 747-400 shown during landing. The tires are 49 inches in diameter. Photo: Roel Kroes

The target rotation for the 747-200 is 12°, with liftoff expected at 10°. At 10°, rotation the tail clearance is 40 inches. Although this may sound like a small margin against tail strike, the plane is gaining speed and lift, and the shock strut lifts the plane, thereby increasing the margin against tail strike. The target climb attitude for the Boeing 747-200 is 14° to 19°, with the lower value used for heavier takeoff weights.

Captain Hedges Explains: When Does the Pilot Transition from Takeoff to Normal Climb-Out?

We typically focus at the far end of the runway until rotation, at which point we start to transition to alternately scanning the instruments and outside until the plane is established in the climb and off the ground. Then we focus largely on the instruments. It's also situational. If you take off with visibility of one-eighth of a mile, you will be glued to the instruments as soon as you are off the ground; if you are in clear weather and there's lots of traffic, you'll spend more time looking outside.

There is a delicate balance between lift and rotation. Trying to rotate at too low a speed reduces the lift and increases the likelihood of a tail strike. Boeing advises that for every 5 knots of speed below the recommended rotation speed, the angle of attack must be increased by 1° (to gain additional lift that compensates for the lower speed). To allow time for the plane to lift, the plane must not rotate too quickly. The target rotational speed is 2°–3° per second, with expected liftoff in three to four seconds.

Additional friction from tail contact can make it more difficult to take off. But the following nonfatal accident demonstrates that a tail strike—even one greater than occurred during the Halifax accident—will not prevent a 747 from taking off.

Singapore Airlines Flight 286, a Boeing 747-400, took off on March 12, 2003, from New Zealand with 389 passengers and crew. Like the Halifax accident, the plane weighed 220,000 lbs more than the value entered into the flight computer. The incorrect rotation speed calculated for the incorrect lower weight was 130 knots, or 33 knots lower than the correct rotation speed. Rotation at too low a speed doesn't allow the shock strut pressure to properly lift the plane and avoid a tail strike. (Both overweight 747 incidents detailed later in this chapter had early rotations and tail strikes.) Singapore 286's tail dragged on the runway for seven seconds (about 1,600 feet longer than the Halifax accident), giving off white smoke.

Being so far from the tail, flight crew often do not notice tail strikes. Sometimes passengers or Flight Attendants in the back of the plane notify an unaware pilot. The latest planes (e.g., Boeing 787 and Airbus A380) have tail strike detections systems—sensors that send a signal to the cockpit displays.

Flight 286's crew noticed a sluggish rotation. The three pilots in the cockpit discussed the possibility of tail strike and concluded it did not happen. The severely abraded tail section indicated otherwise, as did the auxiliary power unit (APU) fire alarm. The APU is a small jet engine in the tail that can power some of the airplane's systems. On the 747, the APU is only used on the ground. Following procedure, the crew discharged the APU fire bottle. The fire alarm continued intermittently.

Because the takeoff weight of an airplane is about 40% fuel, the landing weight assumes a lengthy fuel burn and a much lower weight. The procedure for landing shortly after takeoff calls for dumping fuel to reduce weight, which is not the first choice when flying over metropolitan areas. An overweight landing is not necessarily a safety risk, but it does create an increased risk of structural damage. More concerned about the fire, Flight 286's crew decided to land immediately. The pilot declared an emergency and landed safely (surrounded by fire trucks) just 11 minutes after takeoff.

Because a tail strike can damage the pressure bulkhead (shown in Figure 1.1), the procedure after any suspected tail strike is to not pressurize the plane, get the plane on the ground, and inspect for damage. Without automatic gradual pressurization, the plane must gently gain altitude for passenger comfort, because their ears pop as the plane ascends.

Failure of a pressure bulkhead during a pressurized flight has the potential to destroy the plane. There are two historical examples. In both cases a tail strike caused bulkhead damage, and in both cases the plane landed safely. Faulty repairs (along with metal fatigue) caused catastrophic failure, known as explosive decompression, many years later. Both accidents are reviewed in detail in *Beyond the Black Box*.

Takeoff tail strike impact forces are limited by the forces trying to rotate the plane and therefore have limited potential for plane damage. A takeoff tail strike can lead to a longer takeoff or reduced obstacle clearance, but unless something else bad happens, there is enough extra runway to successfully complete the takeoff. There are no known takeoff tail strikes with disastrous outcomes on the same flight in the modern jet era. The Halifax accident was caused by too little speed and lift; the tail strike was a secondary effect.

Most tail strikes occur during landing. Depending on the plane's motion, a landing tail strike may involve severe impact forces. Tail breakoff during landing, however, is not necessarily as bad as it sounds. In Philadelphia in 1976, for example, wind shear slammed a McDonnell Douglas DC-9 into the runway. The tail was sheared off without any fatal injuries. Today, Doppler radar significantly reduces the threat of wind shear. More recently, in 2013, Asiana Flight 214 broke off its tail landing in San Francisco (see Chapter 9). Although three people died, it is believed that two of the three fatalities could have been prevented by wearing seatbelts.

Stretched Planes

It is common practice for later versions of the same model to be stretched, whereby an additional section of fuselage is added to increase carrying capacity. The Boeing 757-300, 777-300, and 737-800 are all examples of stretched planes; in this case, stretched 23, 31, and 27 feet, respectively, from previous models. The longer fuselage has a greater likelihood of tail strike and requires a smaller rotation angle. There are many other examples, but unique to these three planes is an added tail skid system. The tail skid contains a crushable cartridge designed to absorb energy and protect the tail section. The tail skid system retracts and extends with the landing gear.

Takeoff Acceleration and Newton's Second Law

An incident similar to the Halifax accident occurred on Nippon Cargo Airlines Flight 62 in Japan in 2003. Flight 62, a Boeing 747-200, successfully took off from Tokyo weighing 745,000 lbs, even though the pilot incorrectly entered 550,000 lbs into the flight computer. The plane successfully took off with the following speeds: V_1 = 124 knots, V_R = 132 knots, and V_2 = 146 knots. The correct rotation speed was 167 knots. The Japanese investigators compared the plane's actual acceleration to the acceleration of a 747 weighing 550,000 lbs and one weighing 745,000 lbs.

Acceleration is the rate of change of velocity. Acceleration can also be described as the change in velocity divided by the time interval of the change. A falling apple, when dropped from rest, speeds 32 ft/sec after 1 second, 64 ft/sec after 2 seconds, 96 ft/sec after 3 seconds, and so on. The acceleration occurring over the first 2 seconds is (0 – 64) ft/sec divided by 2 seconds = –32 ft/sec/sec, or –32 ft/sec^2. The negative sign indicates downward acceleration.

The actual Japanese 747, weighing 745,000 lbs, was accelerating at 6.75 ft/sec^2 right before rotation (132 knots). Using $f = ma$, the net forces accelerating the plane can be calculated to be 156,200 lbs. The engines on the Japanese 747-200 were General Electric CF6-50E2 rated at a maximum thrust of 52,500 lbs per engine. The total thrust for the four-engine plane is 210,000 lbs. The difference between the maximum rated thrust of the engines and the actual force accelerating the plane is explained by the forces that resist the plane's forward motion, and by the many factors acting to reduce available thrust.

The thrust available to accelerate the plane is slightly reduced by using engine power for auxiliary equipment (hydraulic pumps, electric generators, etc.). The Japanese 747 also used a reduced takeoff thrust (a common practice) to decrease engine wear.

Airspeed also alters engine thrust during takeoff. There are two competing effects. Engine thrust can be described as the difference between the momentum of the inlet airflow and the outlet exhaust gas. As airspeed increases, the inlet momentum increases; the momentum differential and engine thrust decrease. Greater airspeed also compresses the inlet air, increasing its density; this increases air flow through the engine and increases thrust. During takeoff, the two competing effects initially reduce engine thrust perhaps as much as 20% before it begins to increase again.

The forces opposing the plane's acceleration are air drag and rolling resistance. Air drag is described with a drag coefficient. The drag coefficient has historically been determined experimentally. The drag equation is identical to the previously described lift equation. And just like lift, drag increases with velocity squared. Most rolling resistance occurs in the tires; additional resistance occurs in the bearings. Tire rolling resistance is caused by internal friction in the rubber and by sidewall flexing on each rotation.

Simulations were conducted during the investigation of the incident. The simulator acceleration of the plane weighing 550,000 lbs (at 132 knots) was 8.94 ft/sec^2. Per Newton's Law, the lower-weight plane with the same forces accelerates more. From $f = ma$, the force accelerating the 550,000 lb simulator plane is 152,700 lbs, which is slightly less than the actual plane weighing almost 200,000 lbs more. This slightly lower force is consistent with a slightly less rolling resistance for a plane that weighs less.

Gravitational acceleration from gravity is considerably greater than the acceleration of a 747 taking off, as is the acceleration of a high-end Corvette, which accelerates about four times faster than an underweight 747. The 747 accelerates down the runway more like a poky family car. The pilots have little chance of noticing the difference in acceleration between a somewhat faster family car and a somewhat slower one.

Although it is difficult for the pilot to notice a slower than normal takeoff acceleration, presumably they have a better chance of noticing a sluggish rotation. (Captain Hedges interjects: For sure!) The Japanese 747, trying to rotate at 132 knots instead of the correct 167 knots, experiences less lift than targeted. To compensate, the plane must rotate to a higher angle of attack. The slower speed also reduces the downward horizontal stabilizer force, which in turn slows the rate of rotation. Table 1.1 compares the actual 747 flight to the simulated flight at the correct rotation speed. The actual flight must rotate 30% more, and it takes 60% longer to do so.

As another example, in 2003, a 747-300 took off from South Africa with a weight that was 59% heavier than the value entered into the flight computer (715,300 vs. 448,800 lbs). The pilot delayed rotation by 15 knots after the initial attempt didn't feel right. The Halifax 747 accident plane was 48% heavier.

In 2008, an Airbus A330 took off from Jamaica with an even greater weight error. The plane weighed 74% more than the value entered into the flight computer. Because of a sluggish rotation, the pilot instinctively advanced to full power. As the weight error increases, it becomes easier for pilots to notice rotation anomalies.

Although a weight error has the potential for disaster, it far more likely that the pilot adjusts on the fly—literally. Pilot entry errors were studied worldwide for the period 1989–2009. Thirteen weight errors made it into the safety databases. A few, previously mentioned, had errors similar or even greater than the Halifax accident. Only the Halifax 747 crash involved fatalities. Meanwhile, since 1989, the world has had over a half billion take-offs of Western-manufactured jets.

Table 1.1. Comparison of Actual Japanese 747 Heavy Flight Rotation to Simulated Correct Rotation

	V_R (knots)	Pitch Angle	Rotation Duration (seconds)
Actual plane	132 (too low)	11.9°	7.7
Simulated plane	167 (correct)	9.1°	4.8

Center of Gravity

The center of gravity (CG) is the average location of the plane's weight. The plane's motion can be described as a translation of the CG and a rotation of the plane about its CG. The location of the CG is fundamental to safe flight and must be located within strictly mandated limits. The location of the plane's CG is routinely checked during an accident investigation.

To simplify, the lever calculations associated with longitudinal stability designers of swept-wing planes define an equivalent straight wing known as the mean aerodynamic chord, or MAC. The MAC for the 747-200 is 27.3 feet long. The leading edge of the MAC is located about 7 feet in front of the wing gear (the front set of two main landing gear); the trailing edge of the MAC is about 10 feet behind the body landing gear. The CG limits are given as a percentage of the MAC. Example CG limits for the 747-200 (the limits vary slightly with the plane's weight) are between 13% and 35% of the mean aerodynamic chord, a length of about 6 feet. The CG for the accident plane was 23% MAC, about 10 inches in front of the wing gear. The approximate location of the MAC (and CG) for the Halifax accident plane are shown in Figure 1.12.

The CG will move around depending on how the plane is loaded, and also as fuel is consumed from the inner and outer wing fuel tanks. The CG affects many things on an airplane, including fuel efficiency and handling. The CG also affects rotation. An aft CG makes rotation easier; a forward CG makes rotation more difficult.

Korean Air Flight 520

Korean Air Flight 520 subcontracted cargo loading in Oslo, Norway. On September 21, 2004, the contractor prepared the mass and balance manifest and load plan. Both documents (including calculations) were prepared by hand. The completed manifest, which showed the correct location and weight of all cargo, was faxed to the Korean Air Control Center in South Korea. The certified load master accidentally switched two numbers, and the faulty calculations unbalanced the plane.

The Boeing 747-400 cargo plane took off bound for Seoul, South Korea, a

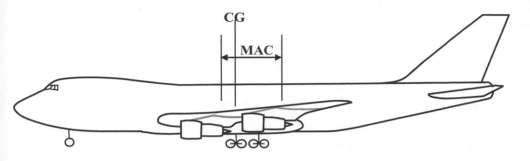

Figure 1.12. Location of the center of gravity (CG) and mean aerodynamic chord (MAC) for the Boeing 747-200.

12-hour flight. The target takeoff speeds were: decision speed V_1 = 143 knots, rotation speed V_R = 154 knots, and takeoff speed V_2 = 166 knots. The plane began to autorotate at 120 knots, which was 34 knots below the target. The pilot lowered the nose and continued the takeoff. The pilot rotated the plane for the second time at the correct rotation speed and took off safely.

En route, the flight crew suspected the plane was misloaded and called Korean Air Operations Control Center in South Korea. The control center verified that the plane's center of gravity (CG) was nearly 16 inches past the mandated aft limit.

The plane's takeoff weight was 725,187 lbs, including 262,500 lbs of fuel. As fuel is burned, the CG routinely shifts farther aft. The projected landing weight of 533,187 lbs moved the CG an additional 20 inches in the wrong direction. During flight, the crew shifted the CG forward about 1 foot by moving pallets. The CG for landing, even after shifting pallets, was still projected to be almost 2 feet past the aft limit.

A tail-heavy plane is susceptible to tail strike (or even loss of control) on takeoff and landing. Boeing recommends aborting the takeoff if the plane self-rotates below the decision speed V_1. During the approach briefing, the flight crew discussed the landing configuration and performance parameters to reduce the probability of tail strike. Meanwhile in Seoul, emergency equipment was placed on standby. After landing uneventfully in South Korea, the nose pitched up at 60 knots when taxiing on the runway. The pilot shut off the engines and parked the plane. The plane was towed off the runway.

The most recent wide-body planes—including the Boeing 747-400, 777, 787, MD-11, Airbus A330, A340, and A380—have load cells built into the axles of the landing gear. This system will also check the center of gravity. Although the system existed on Korean Airlines Flight 520, the crew was not trained in its use—a situation corrected by the airline moving forward. Korean Air also implemented a computerized load planning system that would prevent many data entry errors.

Cargo Planes

Cargo planes are statistically less safe than passenger planes. In one survey, cargo planes in North America were five times more likely to have accidents than passenger planes. The suggested reasons are that cargo planes tend to be older planes with fewer safety features and tend to fly more at night. In another survey, cargo planes were eight times more likely to have weight and balance problems. An extreme example occurred on April 29, 2013, in Afghanistan.

National Airlines Flight 102

National Airlines, operating out of Orlando, Florida, supplies on-demand passenger and cargo service. Employing 230 people (including 43 Captains, 35 First Officers, 13 check airmen, and 21 Flight Attendants), National operated one Boeing 757 passenger plane and three Boeing 747-400 cargo planes. All National cargo planes were contracted by the US Department of Defense to support overseas missions.

National Airlines Flight 102 took off on April 29, 2013, from Camp Bastion, Afghanistan, bound for Dubai, United Arab Emirates, with a stopover for fuel at Bagram Air Base. The plane was loaded with five mine-resistant ambush-protected (MRAP) military vehicles. Depending on the test data referenced, an MRAP vehicle is considered 9 to 14 times more survivable than a Humvee for an improvised explosive device (IED). National Airlines had never before transported heavy MRAP vehicles. In what should have been a red flag for everyone involved, the heavy vehicles shifted during the first leg of the trip: one tie-down strap broke and others became loose.

A simulator filled in performance data for the takeoff out of Bagram. The pilot slowly ramped up to full takeoff thrust in 15 seconds. After 45 seconds, the plane reached a decision speed of $V_1 = 145$ knots. Just 4 seconds later, the plane rotated at $V_R = 145$ knots. Just 54 seconds after the pilot first advanced the throttles, liftoff occurred at 168 knots. Shortly thereafter, things went awry.

As reported by eyewitnesses and recorded on a nearby dashboard camera, the plane was seen to abruptly climb and stall before entering a steep descent. The plane crashed just 30 seconds after takeoff with a (slightly nose-down) violent belly flop. The aircraft immediately burst into a fireball just 590 feet past the runway's end. The front 80% of the plane was consumed by fire. As discussed on page 16, a fireball is related to the plane's sink rate, impact angle, and fuel misting. The sink rate was not recorded because the black boxes failed 3 seconds after takeoff.

The crash killed all seven crew. The Taliban, as is their custom, immediately claimed credit for downing the plane. Military officials were just as quick to deny the claim. Eyewitnesses estimated that the plane only got 500-1,000 feet off the ground.

The crash investigation was initially led by the Republic of Afghanistan with assistance from the National Transportation Safety Board (NTSB) as the representative of the state of the operator and manufacturer. On October 16, 2014, the Afghanistan Civil Aviation Authority delegated the remainder of the investigation to the NTSB.

Two and a half weeks later, the Ministry of Transport and Civil Aviation of Afghanistan reported preliminary findings in a news conference, saying that the vehicles had shifted backward. On May 17, 2013, the Federal Aviation Administration released a "Safety Alert for Operators" reemphasizing standard requirements for transporting heavy vehicles.

It's unclear exactly how many vehicles shifted. (The black boxes mounted in the back of the plane were damaged and failed to record any conversation or data related to loss of control.) Boeing simulated a variety of shifted weight scenarios. In the worst-case scenario, all five vehicles shift backward about 12.5 feet, which in turn shifts the center of gravity about 6 feet aft of the certified limit. Surprisingly, Boeing estimates that the pilots could have controlled the rapid nose-up motion and regained control within five seconds. Something else must have gone wrong.

As explained further in Chapter 3, rotation about the pitch axis (i.e., raising or lowering the nose) is controlled by elevators on the back (trailing) edge of the horizontal stabilizer. Remarkably, the entire horizontal stabilizer moves. The Boeing 747-400 horizontal stabilizer is quite large, 1,470 ft^2, which includes four elevators covering about 22% of the total area. The horizontal stabilizer is moved with a jackscrew powered by a hydraulic actuator—an arrangement used on most large commercial jets.

The 747-400 has four separate hydraulic lines for redundancy and safety. The hydraulic lines raise and lower the landing gear and move the control surfaces. Most control surfaces are controlled by one hydraulic line and a spare. The 747's elevators and hydraulic lines are listed in Table 1.2.

Debris left on the runway (always an important forensic clue), behind the main wreckage, included two small pieces of fuselage skin, a section of hydraulic tubing, a section of a rack support stanchion (the mounting for the black boxes), a bracket for hydraulic tubing, and an antenna from one of the military vehicles on the plane. At least one vehicle had released and damaged the back of the plane. A study of the landing gear and other systems concluded that the no. 1 and no. 2 hydraulic systems failed, no. 4 did not, and no. 3 was indeterminate. The horizontal stabilizer jackscrew was also found fractured, but it was unclear when that occurred. If the jackscrew fractures, the leading edge of the horizontal stabilizer can rotate downward.

Various combinations of vehicle shift and control system damage were evaluated. Assuming that the no. 1 and no. 2 hydraulic lines failed, several combinations of vehicle shift and downward horizontal stabilizer motion were found to destabilize the plane, as summarized in Table 1.3.

The investigators also looked closely at the cargo loading procedures. The plane's takeoff weight of 685,000 lbs (including 207,000 lbs of cargo) was well within the maximum takeoff weight of 870,000 lbs.

The freighter plane had a cargo handling system designed to lock cargo securely to the floor without tie-down straps. But the 18- and 12-ton armor vehicles were too heavy to drive onto the plane, and so the vehicles were mounted on 8-foot by 20-foot aluminum pallets to distribute the heavy load, and slid onto the plane. This configuration could not be secured with the existing floor locking mechanisms, but instead had to be secured with

Table 1.2. Elevators and Hydraulic Lines for the 747

Elevator	Hydraulic Line
Left outboard	no. 1
Left inboard	nos. 1 and 2
Right inboard	nos. 3 and 4
Right outboard	no. 4

Table 1.3. Combinations of Vehicle Shift and Horizontal Stabilizer Motion That Destabilize the Plane

Number of Shifted Vehicles	Horizontal Stabilizer Leading Edge Drops (in)
1	16
2	10
3	3
4	0

Note: Assumes failure of no. 1 and 2 hydraulics.

tie-down straps. Floor locking mechanisms secure cargo for motion in all directions: forward, aft, vertical, and sideways. A tie-down strap is problematic because it pulls in only one direction and is free to move in other directions.

Worse still, National Airlines did not have an established procedure for lashing these heavy vehicles. The 12-ton vehicles were secured with 24 tie-down straps; the 18-ton vehicles with 26. Studies later done by Boeing and Telair (the manufacturer of the cargo handling system) determined one 12-ton vehicle needed 60 tie-down straps to be adequately secured. The National Airlines loadmaster did not correctly identify tie-down attachment points or properly consider effect of strap angle. In fact, because of floor structural limits, Boeing found that only one 12-ton vehicle could have been safely transported. The National Airlines cargo operations manual had not considered transportation of such heavy vehicles.

Passenger Planes with Problematic Centers of Gravity

Passenger planes are less likely to be misloaded (one misplaced passenger is insignificant compared to a shifted 18-ton pallet). Airlines do not precisely weigh passengers and luggage. Instead, they use estimates. To allow for estimation errors, airlines use a more restricted range for center of gravity location. Passenger flights must ensure that passengers are more or less distributed uniformly across the length of the plane.

In 2003, a Boeing 737-800 (with a maximum takeoff weight of 155,500 lbs) took off from Austria with 180 passengers bound for Stockholm with an intermediate stop in Goteborg. In Goteborg, 59 passengers deplaned, and no new passengers got on. The load sheet listed passenger distribution in the four zones as 22, 37, 37, and 25. The pilot reviewed the new and revised load sheet as part of the preflight planning.

Upon taking off in Goteborg, the plane began to autorotate at just 80 knots. The pilot aborted the takeoff. The crew reviewed the load sheet after returning to the terminal. It turned out the actual passenger distribution in the four zones were 5, 26, 54, and 36. The passengers were rearranged, and the plane continued its journey. The investigators later determined that the CG was 10 inches behind the certified limit.

A more extreme example occurred in Rotterdam, Holland, in 2003. The 737-800 was configured for 186 passengers in 31 rows. Only 4 passengers were seated in the first 12 rows. Out of 96 seats, 94 were occupied in the last 16 rows. The pilot barely touched the throttle, and the plane immediately struck its tail. The airplane was taxied back to the terminal and the flight canceled. The plane's center of gravity (CG) was 17.5 inches aft of its legal limit. The CG limits for the 737 must be within an approximate length of 33 inches. A plane will tip over and strike its tail at zero knots if the CG is aft of the main landing gear.

Takeoff Performance Parameters

Originally presented as tables and charts in the flight manuals, the takeoff performance parameters are usually processed today by software. The flight crew inputs a specific runway. Built into the software are the runway's altitude, length, and gradients along the length. Additional inputs are gross weight, location of the plane's center of gravity, and ambient conditions (temperature, wind speed, barometric pressure, and condition of the runway—wet or dry). The outputs are V_1, V_R, V_2, engine thrust, and flap settings.

Captain Hedges Explains: Takeoff from the Best Seat in the House

Takeoff in an airliner is procedurally exacting. Although different airlines have varying procedures, every airline emphasizes strict adherence to published procedures because decision making during the takeoff roll must happen quickly, particularly as an aircraft approaches V_1, a precalculated speed commonly referred to as "decision speed," although that nomenclature can be a bit misleading. V_1 is the fastest speed the aircraft can attain and be guaranteed to stop on the remaining runway; the practical implication is that V_1 is not truly a decision speed, as the decision must be made prior to V_1 as the aircraft is accelerating. A rejected takeoff (or "abort") started at V_1 will see the aircraft exceed V_1 during the abort. Above V_1, the aircraft is guaranteed to be able to safely continue the takeoff even with the failure of one engine, and by regulation will be able to climb to a minimum of 35 feet above the ground by the end of the runway at the takeoff safety speed, V_2. V_1 is thoroughly briefed prior to each takeoff and is discussed in detail in Chapter 2.

All these calculations are predicated on having accurate weight and balance data for the aircraft, as takeoff performance is extremely variable depending on numerous factors, including flap setting, temperature, altitude, runway condition (is it dry, wet, or icy?), runway length, slope, terrain, obstacles in the flightpath, and, most of all, weight. Accurately knowing the weight of an aircraft is critical to ensuring valid takeoff performance data. In a 747, for instance, the aircraft's empty weight may be 400,000 lbs less than its maximum takeoff weight, and the speeds for V_1 can vary from a low of around 110 knots to a high of over 160 knots, a variation of 50 knots or more (the exact numbers depend on what model of 747 it is, what type of engines are installed, and what the maximum weight the particular aircraft is certified for). Along with calculating all speeds necessary for takeoff, pilots also calculate the engine thrust required for every takeoff.

Airlines save maintenance costs by reducing takeoff thrust levels to the minimum required for a given takeoff scenario. Airlines can save hundreds of thousands of dollars per year per aircraft by reducing thrust. The turbine section (or "hot section") of a jet aircraft heats up quickly when an engine is brought up to takeoff power and cools down relatively fast as the aircraft climbs out. On a JT9D engine like MK Flight 747 had, the temperature of the high-pressure turbine reaches approximately 1,075°C. Reducing thrust reduces that temperature, and because increased turbine metal temperature increases engine wear exponentially, it's a huge benefit to airlines to reduce takeoff thrust when possible. But there is a limit: reduce it too much, and the plane will be unable to take off in the runway available.

Gross errors in aircraft weight computations can result in dangerous incidents or accidents. Airlines have specific protocols to calculate weight and balance data, and although manual calculation is possible, it is now normally done by computer. Some airlines use an onboard computer for the purpose, while others do the calculations in a large operations center and electronically uplink the data to the aircraft. Some airlines have now gone so far as to ban the use of manual calculations, the theory being that because it is so rarely required, crews are not especially proficient in reading the arcane and complex charts that are involved. The threat of a simple mathematical error can also go unchecked more easily with manual data, though human error has been found even in extremely regimented systems.

Such was the case with previously discussed Nippon Cargo Airlines Flight 62, a Boeing 747-200 freighter, on October 22, 2003. As the aircraft departed Tokyo, the Captain (who was the Pilot Flying) pulled back on the yoke at the computed rotation speed (V_R) of 132 knots to bring the aircraft's nose up, and he felt the aircraft respond sluggishly and take an abnormally long time to become airborne. At 3,000 feet, air traffic control issued the flight with a clearance to turn right; when the Captain started the turn, the aircraft's stall warning stick shaker activated, advising the pilots that they were dangerously slow (this system measures an abnormally high angle of attack) and close to an aerodynamic stall, a topic discussed in Chapter 4. The Captain then recovered by lowering the nose (which was correct, although increasing power to accelerate more rapidly would have also been appropriate) and asked the Flight Engineer to quickly recompute the aircraft weight. The Flight Engineer realized his error: the aircraft had taken off at 745,000 pounds, but he had inadvertently used the speeds for 550,000 pounds, as that was the aircraft's zero fuel weight, a number he had recently been working with. The Flight Engineer then provided the pilots with the correct speeds, which were 28 knots higher (actual V_R was 168 knots), the pilots reset the markers on their airspeed indicators, accelerated, complied with the new speeds, and discussed the situation. There was concern about a tail strike (hitting the tail of the aircraft on the pavement) because the aircraft had proceeded along the runway in a nose-high attitude owing to rotating at an abnormally low speed. The crew made the wise decision to jettison fuel to lighten the landing weight and returned to Tokyo, where an inspection revealed damage to the tail of the aircraft from scraping the runway.

Although it might seem inconceivable that a crew would make a nearly 200,000 lb error, accidents rarely have a single cause, and that was certainly the case here. There was time pressure on the engineer to maintain schedule, the First Officer was in training, and the takeoff was at night—all human factors issues that increase risk. Cargo airlines in particular fly a lot at night, when circadian rhythm and sleep issues can have a significant impact on alertness and cognitive abilities; unsurprisingly, cargo airlines have a higher accident rate than similar passenger airlines, although it's hard to precisely quantify how much of the accident rate is fatigue related. The First Officer was new to the airline and had previously only flown using kilograms instead of pounds (as Nippon Cargo Airlines did), so the difference in units did not register with him. He also had just completed simulator training and reported that in every simulator training period the instructor used the same weight, 530,000 pounds, so the performance numbers didn't seem unreasonable. The First Officer also made the excellent point that at his previous airline both the Flight Engineer and First Officer independently calculated performance data to ensure gross errors were not made; wisely, Nippon Cargo Airlines changed its procedure after this incident to have both the Flight Engineer and another pilot independently verify the data. It's now accepted industry practice that two people verify performance data, while simulator training is now intentionally conducted at a wide variety of aircraft weights and flap settings.

There are other examples of taking off with incorrect performance data, some with even greater discrepancies than Nippon Cargo Flight 62. Singapore Airlines Flight 286 took off from Auckland about 220,000 pounds heavier than the crew believed, which also resulted in a premature rotation, leading to a major tail strike that damaged the aft fuselage so severely that it triggered intermittent warnings of a fire in the auxiliary power unit, a small turbine located in the tail of the 747 for electrical power and air supply while the plane is on the ground. The crew returned to Auckland, where fire trucks verified there was no fire, and the aircraft was towed to the terminal. Unsurprisingly, the investigation found that incorrectly using data for the lower weight was causal, and once again there were other contributing factors. The Captain had only 54 hours of experience on the 747-400, and lack of flight crew experience in a certain model aircraft increases susceptibility to tail strikes (and errors in general). After the sluggish rotation and liftoff, the crew received the APU fire warning immediately followed by a stick shaker, just as Nippon Cargo Flight 62 did. Also like that flight, the Captain failed to advance engine thrust to maximum, as directed in Singapore's Operations Manual. In the event of a stall or approach to stall conditions (indicated by stick shaker or airframe buffet), a pilot must fly the aircraft first and ensure the stall is broken; part of this process is to increase thrust. Because of incidents and accidents like these and the one involving Air France Flight 447 (see Chapter 4), the industry has greatly increased stall awareness training in the simulator, and the Federal Aviation Administration has passed new rules mandating enhanced stall training and simulator realism. Lack of crew resource management was apparent in the Singapore Flight 286 accident as well, as the third pilot on the flight deck could have independently verified the takeoff data, a common line of defense against errors. This was at the time at the Captain's discretion, and in this case the third pilot said that although he normally did double check the data card, on the accident flight he did not because he was busy dealing with the flight's delay status with the Auckland station manager. This is another example of how accidents have multiple causes, and breaking any one link in the accident chain can prevent the accident.

The crews (and passengers) of Nippon Cargo Flight 62 and Singapore Flight 286 were lucky. They were operating in conditions that allowed their aircraft to successfully become airborne: lengthy runways with a lack of obstacles in the takeoff path. Obviously, a shorter runway or an obstacle would reduce the margin of error; the right combination of a heavy aircraft operating at slow speeds with insufficient thrust and a short runway with an obstacle is a potentially deadly situation. On October 14, 2004, in Halifax, Nova Scotia, MK Airlines Flight 1602, a cargo 747, attempted to take off with incorrect takeoff data, experienced a tail strike, briefly became airborne, struck an earthen berm, and crashed. As in the earlier incidents, the crew used a grossly inaccurate weight for data calculations; in this case, they used data for about 250,000 pounds less than the actual takeoff weight. Unlike the incidents in Tokyo and Auckland, the Halifax runway was considerably shorter, 8,800 feet long with an additional 1,000-foot clearway, which yielded 9,800 feet available for takeoff, 2,126 feet shorter than runway 23L in Auckland and a whopping 3,323 feet less than runway 34L at Tokyo's Narita Airport.

MK Airlines operated under the oversight of the Ghana Civil Aviation Authority, and under the duty time regulations in place at the time, an augmented crew of three pilots with two Flight Engineers was allowed to be on duty for 24 consecutive hours (compared to 18 hours maximum for three-pilot crews based in the United States or United Kingdom). Off-duty crew could rest in the onboard crew rest area. At the time of the accident, the flight crew had been on duty 19 hours, and their day was scheduled to be 24.5 hours, 30 minutes more than allowed. Because of delays, if the trip had continued safely it would have required 30 hours on duty. Also on board was a ground engineer (a maintenance professional) and a loadmaster, the person responsible for correctly loading, distributing, and securing the cargo. At the time of the accident, the ground engineer and loadmaster had been on duty for 45.5 hours, and there were times that they could spend up to seven days on board an aircraft.

Until recently, fatigue has been rarely cited as a contributor to aircraft accidents because it is difficult to quantify. After much academic study and several high-profile accidents—notably, the loss of Colgan Flight 3407 in Buffalo, New York, on February 12, 2009—fatigue garnered widespread attention in aviation safety and regulatory circles. The Colgan accident brought about an enormous restructuring of flight and duty time limits for pilots in the United States.

MK Flight 1602 was part of a long day of flying. Beginning in Luxembourg, the flight flew to Bradley International Airport (serving Hartford, Connecticut) and then departed for Halifax. After Halifax, it was to fly to Zaragoza, Spain, and finally back to Luxembourg in one big circuit. Owing to multiple delays both in Luxembourg and at Bradley, takeoff in Halifax didn't commence until 3:53 a.m., a time close to the nighttime circadian trough, where people are most fatigued. This trip was a virtual prescription for jet lag: the long duty periods, lack of sleep, and "back side of the clock" nighttime flying made it more important that the pilots adhere to procedures, become more methodical, and express their concerns more clearly than ever. Unfortunately, increased fatigue frequently leads to errors, some of which are subtle.

Recall that most airlines now do weight and balance and takeoff performance calculations with a computerized program. That was not always so, and MK Airlines had recently converted to a computerized system called the Boeing Laptop Tool (BLT). The crews were given a 46-page manual but no formal training: self-study was the only way to become proficient. Unfortunately, with any new and complicated system, there are opportunities for errors; in highly automated systems these can be quite subtle. In this case, investigators believe that the user of the BLT was not fully conversant on its features and was unaware that under certain circumstances a reversion feature in the BLT software could allow propagation of data from the previous flight. In fact, the data used in Halifax were nearly identical to the data used in Bradley. Going from Halifax to Spain required much more fuel than the flight from Bradley to Halifax, which only took an hour and nine minutes; combined with a full load of cargo, the airplane was much heavier, needed much faster speeds to fly, and required higher takeoff thrust as well. Ironically, the computerized system that was supposed to make performance calculations easier was contributory to the crash.

Another difference between the Halifax crash and the earlier incidents was the presence of the earthen berm of the end of the runway. This berm was installed to mount a navigational aid and was vetted by Canadian au-

Flying with Fatigue

How tired is too tired? That's the question researchers from Australia and New Zealand addressed in 2000. Their experiment evaluated a sample of truck drivers and military personnel in a series of tests on two separate nights, once after being sleep deprived for up to 28 hours, and once after drinking alcohol. The results of the two tests were remarkably similar: after 17 to 19 hours without sleep, the subjects did as well as subjects with a 0.05% blood alcohol concentration (BAC); their response speeds were up to 50% slower than well-rested or completely sober people. Longer periods of sleep deprivation made judgment, accuracy, and reaction times even worse, equivalent to someone with a BAC of 0.1%, which is legally drunk. The safety significance is clear: fatigue is an important factor in industrial and transportation safety.

thorities as safe because it met all requirements for aircraft clearance. Aircraft are required to attain at least 35 feet of altitude by the end of the runway and this berm was well below that, so it was not considered an obstacle for takeoff calculation purposes. Ultimately, if the berm had not been there, the aircraft would have successfully completed its takeoff, though it would have been damaged by the tail strike. Investigators urged airport operators to conduct a hazard analysis on manmade objects off runway ends and to mitigate hazards when possible, with solutions like frangible structures designed to break away when struck by an aircraft. On June 24, 1975, Eastern Flight 66 crashed during a wind shear encounter at New York's John F. Kennedy International Airport. Contributing to the severity of the accident was that the Boeing 727 first contacted rigid (nonfrangible) metal runway approach light towers, which severely damaged the aircraft. After that accident, frangible structures near runways became the norm in the United States and much of the world.

Lack of adherence to procedures, lack of effective oversight from the Ghanaian authorities, extremely long crew duty days, flying late at night with little sleep, and lack of effective crew resource management added up to an accident waiting to happen. Experience helps pilots overcome some errors, and in this case the crew was extremely experienced: the Captain had 23,200 hours of flying time, and the First Officer had 8,537. That these were seasoned pilots is not in doubt; that cumulative factors allowed this crew to have an accident was symptomatic of a greater problem in the safety culture within the company. The biggest changes MK undertook after the accident were organizational, with a host of qualified safety professionals joining the company. In addition, a safety culture questionnaire was sent to employees, frontline staff feedback was solicited, and crew scheduling changes were made in an effort to combat fatigue.

Weight and balance are fundamental concepts for safe flight. It's clear why the airplane can't be overweight for takeoff and why the correct speeds must be used, but balance is just as important. A plane can be thought of as a fulcrum, and it balances at its center of gravity. Every plane has a center of gravity (CG) envelope it must stay within to ensure safe flight. There are forward and aft limits for the CG, and a typical CG in most airliners would be around 25% mean aerodynamic chord. Pilots must ensure that the CG is within limits for the entire flight, as it moves with fuel burn in swept-wing aircraft; this is particularly an issue in aircraft like the Boeing 747-400, which has a fuel tank in the horizontal stabilizer, as failing to burn that fuel on the correct schedule can lead to a CG that is out of limits. In a jet airliner, the CG will always be forward of the main landing gear, or it will sit on its tail on the ground. This is a factor when servicing some longer aircraft, as ground personnel must be careful to load cargo in the front of the plane before the back, and unload the back cargo before the front or the plane will tip; some planes (e.g., the Boeing 737-900) have tail stands that are removed before flight to ensure this doesn't happen.

While the ground handling personnel are versed on loading and unloading the aircraft, the pilots concern themselves with ensuring the CG is safe for the planned flight. If an aircraft takes off with a CG that is too far forward, rotating the aircraft will require more force, and if the CG is extremely far forward, it may not become airborne at all. In the air, a forward CG makes the aircraft generally more stable, though on landing the nose may de-rotate and firmly contact the runway.

A more serious state is an excessively aft CG. In this situation, the aircraft may spontaneously rotate at a much slower than normal speed. On September 21, 2004, just three weeks before the Halifax crash, a Korean Air cargo 747-400 departed Oslo, Norway, with an incorrectly loaded aircraft. As seen in the other accidents and incidents, errors were made in the preparation of load plan and were not caught. When the flight departed, the CG was at 37.8% MAC, while the aft CG limit was 33% MAC. Although the planned V_R speed was 154 knots, the aircraft began to rotate by itself at 120 knots. The V_1 speed for this takeoff was 143 knots, but the Captain elected to continue the takeoff even though this is a critical situation where an abort would have been a reasonable decision. Korean Air's training program emphasized that high-speed aborts are generally a worse option than continuing the takeoff and sorting out the problem airborne. This training is statistically sound, but sometimes an abort is the best decision. In this case, the Captain could have justified an abort because he could not be certain the aircraft was capable of being flown safely. How could he know whether the pitch would continue to increase, possibly uncontrollably? He couldn't. Boeing commented to the accident board that "Aborting the takeoff should be considered a viable option if a clear case of autorotation is perceived prior to V_1."[12] Fortunately, through skilled handling, there was no tail strike, and after conferring with Korean Air operations, it was determined that the CG was too far aft and would continue to move farther aft as fuel was burned. The decision was made to shift some pallets in cruise to minimize the aft CG and continue to Seoul. The aircraft landed safely and was towed to parking.

Although the Korean incident resulted in no damage to the plane, there are numerous accidents associated with incorrect loading. On January 8, 2003, Air Midwest Flight 5481, a Beech 1900D, took off from Charlotte, North Carolina, with an actual CG of 45.5% MAC, while the farthest aft allowed is 40% MAC. The crew was unaware of this, as they had received data indicating the MAC was at 37.8% MAC, within the limits. There were numerous errors made in the accident chain, but the most serious involved maintenance performed on the aircraft two nights earlier, part of which involved inspecting the elevator control cables. The rigging of flight controls is a precise task, and it was new to the mechanic doing the work. The aircraft was incorrectly rigged, yet it successfully flew nine flights before the accident flight. What made the accident flight different? The aft CG. When the aircraft rotated, pitch kept increasing. Because she could not arrest the pitch rate, the Captain asked the First Officer to help her push forward on the yoke in an attempt to control the aircraft. Unknown to them, the incorrectly rigged controls gave them only 7° of nose-down elevator travel, half the normal amount. The aircraft pitched up steeply, rolled abruptly to the left, stalled, and crashed into a maintenance hangar. Once this aircraft became airborne, there was no way that the crew could have possibly controlled it. Even if cargo starts out in the correct place, if it is improperly secured, it can shift in flight with disastrous consequences, as previously described in the Bagram accident.

CG management is critical to aviation safety, and most of the hazard mitigation is done on the ground long before the engines are started. Crews must be aware of the hazards of using correct weights for takeoff data, ensuring that accurate calculation of CG, takeoff speeds, and thrust settings are used. In some operations, the crew has limited insight into the raw data used to formulate performance data, but in all cases a "reasonableness check" should be made, where pilots mentally add up approximately how much the plane should weigh, including passengers, cargo, and fuel, and cross-checking this with available data and their prior experiences with that aircraft. Good procedural design has now made these accidents rarer, and all companies should have a minimum of two people checking critical information. Training has become much more focused on proactively identifying errors and mitigating hazards through adherence to checklists and procedures. One of the most important things a Captain can do is to form an effective team with the other crewmembers by encouraging subordinates to speak up if something doesn't look right or if they have a concern.

2

Takeoff (Never Mind!)

Kalitta Air began operations in 1967 with a single Cessna that flew critical assembly-line components to Ford Motors. After a series of mergers, bankruptcies, and buyouts, Kalitta reemerged in 2000 flying cargo with three Boeing 747s. On the night of September 11, 2001, a Kalitta Air 747 (hauling supplies from the West Coast to disaster workers at Ground Zero in New York) was the only non–US Air Force fighter plane in the sky.

By 2009, Kalitta Air operated 18 aircraft; mostly older 747-200s and two newer 747-400s. Kalitta flies international scheduled and ad hoc cargo charter services, including the following route: New York City–Brussels–Bahrain–Dubai–Brussels–New York City. To fly international flights with relief crews, Kalitta had 1,486 employees, including 364 flight crew and 425 mechanics.

On May 25, 2008, a Kalitta Air Boeing 747-200F was cleared for takeoff at 11:29 a.m. local time in Brussels. An older plane, it was flown by a Captain, First Officer, and Flight Engineer. (All modern digital jets fly with two pilots.) The plane accelerated to V_1 (138 knots) in approximately 44 seconds. Upon reaching V_1, the Captain transferred his right hand from the throttles to the control column to prepare for rotation. The decision speed V_1 is the last possible instance to safely reject the takeoff. At V_1, Boeing estimates the aircraft had used up 5,290 feet of the 9,800-foot-long runway. About 4 seconds later, the cockpit voice recorder recorded a loud bang:

11:30:46 BANG!

11:30:46 Flight Engineer: "Whoa."

11:30:48 Captain: "Reject."

11:30:51 First Officer: "Tower Connie eight zero seven heavy—two zero seven rejecting runway two zero."

11:30:55 Control tower: "Connie two zero seven roger; can you vacate to the right?"

11:30:58 Captain: "Negative."

11:30:59 Copilot to tower: "Negative."

11:30:59 Flight Engineer: "Negative, negative, negative."

11:31:00 Captain: "We're gonna take the overrun."

11:31:01 First Officer to tower: "We're taking the overrun."

11:31:02 Control tower: "Roger."

11:31:04 Sound of impact and metallic grinding.

The pilot later stated that he thought the plane was no longer accelerating and aborted the takeoff by transferring his hand from the control column back to the throttles. Within two seconds, the engine throttles for all four engines were brought back to idle, but not before the plane reached a maximum speed of 152 knots. Because of the engine's rotational inertia, there is a delay between throttle reduction and thrust reduction (two seconds for a 70% thrust reduction is a realistic number). At maximum speed, Boeing estimates that there was 3,434 feet of remaining runway. Thirteen seconds later, the plane departed the runway at 72 knots. The aborted takeoff is shown approximately to scale in Figure 2.1.

After leaving the runway; the plane traveled 738 feet across a field, dropped off a 13-foot embankment, and broke into three sections. The un-

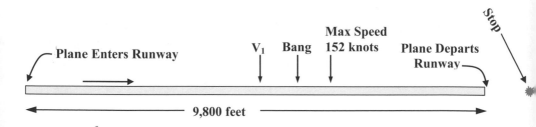

Figure 2.1. Events on the runway involving the Kalitta Air accident plane drawn approximately to scale.

injured flight crew safely departed the plane. The nose gear collapsed and pushed into the fuselage. The main landing gear consists of left and right wing gear and body gear. The main landing gear attach to the fuselage with a wing spar and gear beam. The left wing gear separated from the wing spar and pushed through the upper wing; the right wing gear separated from the wing spar. The left body gear remained attached; the right body gear fractured at the wheels. The extent of the damage is shown in Figure 2.2.

The fire brigade arrived within the prescribed three minutes and scanned the wreckage with thermal cameras, as did the federal police with a flyover. There were small fuel leaks, but no fire. Although unpredictable things can happen at the fracture lines, this accident, if it were a passenger plane, would be survivable.

The US National Transportation Safety Board[1] (NTSB) participated in the investigation as the representative of the country of origin for the aircraft manufacturer and airline. The NTSB also provided support for readout of the voice and data recorders and specialists for the various reports (i.e., structures, engines, airplane systems, etc.). Also participating in the investigation were the Federal Aviation Administration (FAA), Boeing, Pratt & Whitney (manufacturers of the engines), and Kalitta Air.

Figure 2.2. The accident plane fractured into three sections. Wikimedia/Bladt

What Caused the Bang?

To understand the loud boom, we must understand how jet engines operate. We must also understand reverse thrust, one of several devices available for slowing the plane.

The simplest model of a jet engine is a balloon. If the nozzle is pinched shut, the pressure is balanced in all directions. If the nozzle is open, the pressure cannot push against the opening and the pressure on the opposite surface creates a forward thrust, as shown in Figure 2.3. The balloon rapidly loses pressure. A turbojet is a container filled with turbomachinery designed to maintain a constant pressure.

Turbojet

A compressor supplies pressurized air to the combustion chamber. The compressor adds energy to the airflow by forcing the air into progressively smaller volumes with multiple stages of compression. Each stage consists of a set of fixed blades and a set of blades attached to a rotating disk. The fixed blades redirect the airflow to create the optimum angle of attack for airflow across the rotating blades. The added energy increases the pressure and temperature of the air as it progresses through the various stages of the compressor. The first World War II–era German turbojet compressor had a

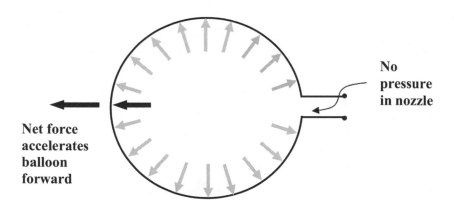

Figure 2.3. With an open nozzle, unbalanced pressure results in forward thrust.

pressure ratio of 2.8; that is, the compressor increased the pressure of the air to 2.8 × 14.7 = 41 psi.

In a turbojet, fuel sprayed into the combustor is ignited. Rapid expansion of the hot combustion gases maintains the pressure despite massive outflows blasting into the turbine section. Instead of adding energy to the air, the turbine section extracts energy like a windmill to rotate a shaft. The other end of the shaft powers the compressor. The hot exhaust gases lose pressure, temperature, and energy as they pass through increasingly larger volumes in each stage of the turbine.

All the strokes of a piston engine are present: intake, compression, combustion, and exhaust, albeit continuously instead of individual discrete motions, which is a major advantage of jet engines. The smooth, continuous motion leads to less wear and tear, fewer parts in the drive train, and a dramatic increase in reliability.

A piston engine starts with a few cranks from a starter. A turbojet needs significantly more power and requires a continuous minimum shaft speed for successful ignition. It's like a car not starting until the engine is rotating continuously at 400 RPM.

Within limits, the turbomachinery is self-correcting. Increased fuel flow increases the amount of exhaust gas, which in turn rotates the turbine faster. The faster turbine rotates the compressor faster, thereby providing the additional air required to burn more fuel.

Turbojets, however, are optimized for a specific set of conditions (i.e., power, altitude, and RPM). Straying too far from optimal conditions and the first blades on the early stages can result in a stall (the blades act like rotating wings and can stall like an ordinary wing). The last stages of the compressor can turbine. Turbining compressor blades act like turbine blades, extracting (instead of adding) energy to the airflow. This can lead to compressor surge or flow in the wrong direction—the eventual topic of this section.

To increase operating flexibility, a two-spool design was developed. A spool is a section of compressor and turbine that share the same shaft and rotate at the same RPM. The second spool is essentially an outer shaft that covers an inner shaft. The limitations of a single spool are reduced (not eliminated) by operating the two spools at a variety of speeds. Rolls Royce uses a three-spool design, with three shafts rotating at three different speeds.

The first two-spool turbojet, the J57 (thrust 10,500 lbs), was designed by Pratt & Whitney for the Boeing B-52, an eight-engine US Air Force bomber. A commercial version, the JT3, was used on the Boeing 707, the first successful jet airliner. The J57 is shown schematically in Figure 2.4. The inner spool has nine stages of low-pressure compressor connected to two stages of low-pressure turbine. The outer spool has seven stages of high-pressure compressor connected to a single stage of high-pressure turbine. The first J57 had a pressure ratio of 11 (162 psi). For each new jet engine design, the pressure ratio increases, albeit slowly.

The drive train of a jet engine is simple, as Figure 2.4 shows. A cutaway of an actual jet engine, however, reveals quite a bit of complexity. Additional engine systems weighing up to half of the powertrain include: auxiliary power (an attached gearbox), ignition system, bleed air (for anti-icing systems, internal cabin pressurization, engine control, and turbine blade cooling), hydraulic pumps, fuel pump lubrication system (for shaft bearings), and filters. To provide additional operating flexibility, the fixed compressor and turbine blades are adjustable. Hydromechanical and electrical systems are required to monitor and control all the above.

Speed is another advantage of the turbojet. The speed of sound is a practical limit on propeller tip speed. At speeds above (or approaching) the

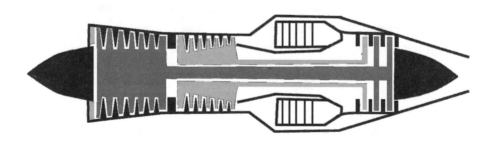

Figure 2.4. Schematic of Pratt & Whitney's J57, the first two-spool turbojet developed for the B-52.

speed of sound, the propeller is moving faster than the air can get out of the way. Energy is wasted, compressing the air instead of creating thrust.

As described in Chapter 1, a thrust force (or a lift force) can be described as a flow rate—pounds per hour × speed of the jet exhaust. The turbojet exhaust is also limited by the speed of sound. The hot exhaust gases can travel much faster than the propeller because the speed of sound[2] increases with temperature. The Boeing 707 circa 1958 cruised at 495 knots, about 200 knots faster than Boeing's final propeller plane.

This faster speed came at a price. The turbojet's high-speed exhaust was extremely noisy. In fact, jet engines were banned from all airports in metropolitan New York City until they became quieter.

Turbofan

Thrust can be increased by increasing the flow rate or increasing the speed or both. But the turbojet's exhaust gas contains wasted energy, described by $\frac{1}{2}$ × mass × velocity^2. Less energy is wasted if a larger quantity of gas is moved at a slower speed.

A turbofan is a turbojet with a ducted fan in the front. The fan blades create additional thrust by blowing air—which bypasses the turbojet core —backward like a propeller. Engineers discovered that additional turbine blades connected to fan blades in the front increased efficiency by moving larger quantities of air at lower speeds. Extracting more energy out of the exhaust gas had the added benefit of lowering the exhaust gas speed and creating a quieter jet engine.

The Pratt & Whitney JT9D-7Q (53,000 lbs thrust) turbofan engine used on the Kalitta Boeing 747-200 accident plane is shown in Figure 2.5. Continuing the trend of increasing thrust and pressure ratio, this circa 1970s engine had a pressure ratio of 25; the ambient air is pressurized by the compressor to 25 × 14.7 psi = 367.5 psi. Each generation of jet engine has a higher-pressure ratio. The pressure ratio for the latest Boeing 787 engines (first commercial flight 2009) is 47.

The JT9D-7Q has 46 fan blades. The fan, low-pressure compressor, and low-pressure turbine section are all connected to the same shaft and rotate at a speed designated as N1. The high-pressure compressor and high-

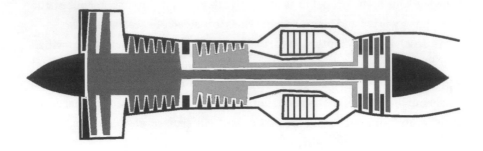

Figure 2.5. Schematic of Pratt & Whitney JT9D-7Q turbofan engine.

pressure turbine section share a shaft and rotate at a speed designated as N2. N1 and N2 for the JT9D-7Q engines are about 3,600 RPM and 7,800 RPM respectively. Rotating at such extreme speed creates the possibility of the engine ripping itself apart with centrifugal forces.[3]

In a turbojet, all the inlet air passes through the engine. In a turbofan, most of the fan air bypasses the turbojet core and creates most of the thrust. The larger-diameter turbofan cowling and smaller-diameter inner core turbojet are shown in Figure 2.6.

The bypass ratio for the JT9D-7Q is 5 to 1; that is, 5 lbs of fan air bypass the inner engine for every pound of air that goes through the turbojet core. The larger the bypass and pressure ratios, the greater the amount of energy extracted from the exhaust gases. Just like the pressure ratio, the bypass ratio keeps increasing incrementally with each new design.[4] The bypass ratio of the newer Boeing 787 engines is about 9.

If air does not flow properly in the front of the compressor, the higher-pressure air in the back of the compressor can flow in the wrong direction with a loud bang, a condition known as a compressor surge. Airflow through the compressor can be disrupted by engine deterioration, engine control malfunction, a crosswind, or by ingesting a variety of things. The history of jet engines is filled with incidents involving compressor surge. In modern planes, compressor surge most commonly (but not always) self-corrects.

A toy balloon pricked with a pin pops when it releases a small volume of low-pressure air (less than 1 psi). One can only imagine the loud bang associated with a much larger volume of air pressurized to over 360 psi. Pilots who have experienced a compressor surge report that it can be as loud as a shotgun blast a few yards away. Other possible signs of a compressor surge: more than one loud bang, instant loss of thrust that yaws the plane, fluctuating engine parameters, and visible flames from either end of the engine.

The Kalitta Air 747 engine number 3 was disassembled for inspection. Organic debris was found on five fan blades and on each stage of the low-pressure compressors. No engine damage was found. Samples of the organic residue were sent to the Smithsonian Institution's Museum of Natural History for DNA analysis. The DNA was identified as belonging to the European kestrel, a small bird of prey (typically weighing 5 to 6 ounces) in the falcon family. Brussels Airport uses a variety of means for dispersing birds, including noise and lethal methods. The kestrel, however, is protected by law—lethal methods of dispersal are illegal.

Figure 2.6. Pratt & Whitney JT9D for the Boeing 747-100, the first model 747. Note the larger-diameter front fan cowling and smaller turbojet core. Wikimedia/Cleynen

As part of the certification process, engines are tested for bird ingestion. To pass the test, the engine must ingest four 2.5 lb birds with no more than a 25% temporary loss of thrust. As for the Kalitta 747, a slight drop in engine pressure was recorded on the flight data recorder. With data recorded only every four seconds, a larger drop could have been missed.

The certification testing also requires ingesting an 8 lb bird, an event that destroys the engine. To pass the test, the engine must safely shut down without damaging the plane or ejecting material through the engine casing. The concerns are rotating unbalanced weight and fire. US Airways Flight 1549,[5] an Airbus A320, famously ditched in the Hudson River after ingesting at least two Canada geese (about 8 lbs each) in both engines.

Rejected Takeoff

The concept of rejected takeoff (or RTO, also called an "aborted takeoff" or "abort"), when the pilot changes his or her mind about takeoff partway down the runway, is fundamental to a safe departure. To allow the best decision with the briefest possible delay, all pilots must be 100% focused on the possibility of RTO during every takeoff. Historically, 1 out of every 3,000 takeoffs are aborted, but only 2% occur at speeds higher than 120 knots.

The possibility of RTO is central to the calculations for a safe takeoff distance. Recall the three speeds V_1, V_R, and V_2. If the takeoff is rejected below V_1, there is, by design, sufficient runway to safely stop the plane. Or if an engine fails above V_1, the runway is long enough for the underpowered plane to complete a safe takeoff, obtaining an altitude of 35 feet by the end of the runway. If the takeoff is rejected above V_1, there may not be enough runway to stop safely. All airliners are certified to complete the takeoff if one engine fails above V_1, For that reason, the safest course of action for the Kalitta Air 747 was to complete the takeoff.

For the conditions of the accident flight (wet runway, 692,830 lbs at takeoff, and virtually no wind) V_1, V_R, and V_2 were calculated to be 138 knots, 157 knots, and 167 knots, respectively. The plane is rated for a maximum takeoff weight of 820,000 lbs. The stop margin was calculated to be 897 feet. Stop margin is the distance remaining after the aircraft comes to a complete stop, measured from the nose gear to the end of the available runway. The calculated stop margin assumes that braking begins at V_1. The accident plane, however, entered runway 20 at intersection B1, which was 1,023 feet from the north end of runway 20. From intersection B1, there is 8,776 feet of remaining runway.

The flight manuals state that a takeoff should be aborted if the aircraft is "unsafe" or "unable" to fly. But the manuals do not define unsafe or unable. Investigators of a different incident asked the airplane manufacturer to define those terms. The response: unsafe to fly are circumstances where rejecting the takeoff carries significantly less risk than flying the aircraft. Unable to fly are circumstances where there is a reasonable probability of not being able to control the aircraft if the takeoff is continued. Obviously, "unsafe" and "unable" are subject to pilot interpretation and judgment. Additional definitions were not given because the manufacturer believed it could lead to further misunderstanding and "incorrect decision making."[6] In other words, to explain all the nuances, the manufacturer must write a lengthy and complicated engineering report—too much information for pilots to process within a few critical seconds during takeoff. The combination of circumstances is endless. Clear examples of unsafe and unable follow. Unable: the center of gravity is too far forward, prohibiting the plane from rotating and becoming airborne; a flock of birds causes complete engine failure. Unsafe: the flaps and slats are not properly set for takeoff (a takeoff warning alarm should sound).

There is a long list of events that are considered unsafe for flight and can result in a rejected takeoff, including a center of gravity (CG) that exceeds the specified range (this affects handling; note that this is different than the plane being unable to fly if the CG is so far forward the plane cannot rotate), engine failure, any fire warning, takeoff configuration alarm (i.e., incorrect flaps), air traffic control directives, crossing aircraft, vehicles on the runway, bird strike, and directional control problems. One-off unique events have also occurred, such as a cockpit window opening after V_1. There has also been a long list of alarms or warnings that can occur at the worst possible moment. To provide fewer distractions, digitally controlled planes developed in the 1980s incorporated internal logic that temporarily shut off all but the most critical warnings. One common arrangement shuts off most alarms above 80 knots, and turns them back on when the plane reaches an altitude of 400 feet. Captain Hedges notes: These are some of the things that historically have caused rejected takeoffs (RTOs), not a list of circumstances under which it's always appropriate to abort the takeoff. Tire failure and a cockpit window opening during the takeoff are highly distracting, but are generally safer to deal with once airborne (a blown tire invalidates an aircraft's certificated stopping distance).

Continuing a takeoff with faulty or damaged equipment defies common sense in that a defective airplane seems fundamentally safer on the ground—the only exception is a faulty airplane on the ground at speeds above V_1 with a rapidly approaching end of runway. Studies have shown that most events occurring at speeds higher than V_1 are safer to solve in flight than on the ground. Pilots must fight the uncertainty of a tense situation and the normal instinct to abort the takeoff, and a great deal of simulator training helps them develop the proper RTO decision-making skills.

Required Takeoff Distance

The distance mandated for a safe takeoff is the longest of the following three conditions: (1) the all-engine takeoff, (2) engine failure partway down the runway at V_1 with the takeoff continued, and (3) the distance required to safely stop the plane after a rejected takeoff beginning no later than V_1.

The all-engine takeoff is the most common. This distance is the length required for the lowest part of the plane (the bottom of the landing gear) to reach a height of 35 feet above the departure end of the runway. To be conservative, the regulations add an additional 15%.

The procedure for engine failure above decision speed V_1 is continued takeoff. Even with less thrust the plane continues to accelerate, albeit at a lower rate. The required takeoff distance for engine failure above V_1 is the distance required to meet the 35-foot height rule without the additional 15% margin.

The required distance for a rejected takeoff is the distance it takes the plane to accelerate up to the decision speed V_1 and the additional distance required to safely stop the plane. Circumstances vary, but generally two engine planes (losing half their thrust) are limited by the engine-out scenario. Four-engine planes are often limited by the 115% takeoff distance requirement.

Because runways at a given airport are of fixed length, the takeoff calculation becomes what is the maximum allowable takeoff weight that meets all the safety rules for required takeoff length described above. Obviously, a heavier plane requires a greater distance to meet all the requirements. Before computers, the flight crew would enter a chart in the flight manual that adjusts for runway slope and wind (as well as a few other factors) to calculate a corrected takeoff distance. A second chart was then used to read off a maximum takeoff weight.

For any given weight, a faster plane requires a longer distance to safely stop. Because V_1 is the highest possible speed to begin braking, a higher V_1 requires a longer stopping distance, as shown schematically in Figure 2.7A. Engine failure (or any other failure) above V_1 requires continued takeoff. The distance required to complete the takeoff after engine failure decreases as failure occurs at higher speeds. The relationship between required takeoff distances after engine failure is shown in Figure 2.7B. (There will be different curves for different weights.) The intersection of the two curves, shown in Figure 2.7C, is the so-called balanced field length, which gives the V_1 speed for the shortest possible takeoff distance that meets all safety re-

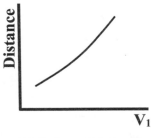

(A) Rejected Takeoff

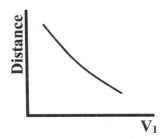

(B) Engine Out and Continued Takeoff

Figure 2.7. Takeoff distance versus V_1 for A, rejected takeoff, B, engine out and continued takeoff, and C, balanced field length.

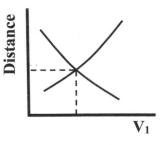

(C) Balanced Field Length

quirements. The manufacturer must establish V_1 speed for all possible take-off weights and runway conditions.

Forces Slowing the Kalitta 747

The maximum deceleration of the Kalitta 747 during the rejected takeoff was 0.38 G's, or 0.38 × 32.2 ft/sec² = 12.2 ft/sec². This occurred early in the event after braking forces quickly ramped up and when the plane's speed and air drag were at a maximum. Per $f = ma$, the maximum force stopping the plane is 38% of the plane's weight, or 0.38 × 692,800 lbs = 263,300 lbs. Boeing reports that the wheel braking coefficient was 0.28; the wheel braking force was 28% of the plane's weight, or 182,400 lbs. The difference between the two (263,300 lbs – 182,400 lbs = 80,900 lbs) is the stopping force available from everything else. This includes air drag, runway slope, rolling resistance, and possibly spoilers or speedbrakes.

Spoilers

The spoilers (a.k.a. speedbrakes) also add braking forces. The spoilers for the Airbus A320 are shown in Figure 2.8.

In the Kalitta accident, the pilot said he applied the spoilers, but the spoilers (and the spoiler lever) were found in the retract, or stowed, position. The investigators could not conclusively determine whether the spoilers had been deployed.

The spoilers create two effects. Spoilers create more aerodynamic drag and reduce lift on the wing. Depending on the flap settings, the spoilers may reduce, eliminate, or even reverse lift on the wings. (If lift is the result of air deflected downward, air deflected upward by the spoilers can create negative lift.) The force on the wheels increases, perhaps even exceeding the plane's weight. The reduction of lift (and negative lift) transfers more weight to the wheels and makes wheel braking more effective. Boeing analyzed the effect of the spoilers between 70 and 120 knots. At 120 knots, the spoilers only added a deceleration of approximately 0.5 ft/sec²; suggesting a potential spoiler braking force of 10,700 lbs. Boeing estimated that stopping on a wet runway from 152 knots with spoilers (and no thrust reversers) would have shortened the stopping distance by 731 feet.

Figure 2.8. Spoilers for Airbus A320. Also shown are hydraulic actuators and controls for the flaps and spoilers. Wikimedia/Anonymous

Runway Slope

The runway's slope increased from 0.62% to 0.93% at the point of the engine anomaly. If the geometry is worked out, a 0.93% upward runway slope represents a force of 0.93% of the plane's weight, or about 6,400 lbs of force opposing the plane's motion. This sudden additional force slowing the plane may have contributed to the pilot's impression about the plane's ability to fly. The component of weight that opposes the planes motion is shown on an exaggerated runway slope in Figure 2.9. Obviously, a downward sloping runway makes a plane more difficult to stop.

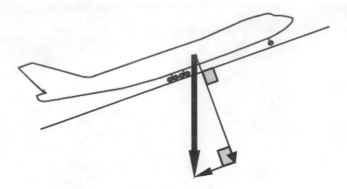

Figure 2.9. Exaggerated runway slope illustrates the component of weight that opposes the plane's motion. Gray rectangles identify 90° angles. The plane's weight is a force that always acts straight down. The components of this force can be resolved into components that are parallel and perpendicular to the runway.

Rolling Resistance

Consider a wheelbarrow, with air-filled rubber tires, being pushed at constant speed. According to $f = ma$, the pushing force should result in constant acceleration, unless there is an equal and opposite force pushing back. The opposing force is the rolling resistance. Most of the rolling resistance is in the tires, but the bearings also contribute. Starting with the definition of work = force × distance, the tire absorbs work and energy every rotation when the tire contact point deforms into a flat spot. The rotating tire continuously requires additional work to flatten a new spot on the tire. The energy dissipated by rolling reappears as the energy or work required to push the wheel barrel (work = rolling resistance × distance pushed). The same principle applies to the tires on automobiles. Noticeably worse gas mileage occurs with underinflated tires. Rolling resistance data for a typical airliner is approximately 1.5% of the plane's weight.

Thrust Reversers

To slow the plane during landing, some propeller planes create reverse thrust by altering the blade angle and blowing air in the opposite direction. A thrust reverser on a turbojet is a blocking device that swings into the jet engine's exhaust and deflects the exhaust gas forward. In a turbofan, most of the thrust is created by the fan blades up front. Blocker doors extend into the fan stream to block the normal air path, and simultaneously the part of the turbofan cowling slides back, exposing slots for airflow. (The back portion of the forward cowl of the JT9D engine on a 747 as seen in Figure 2.6 moves aft when in reverse. It is seen here in the forward thrust position.) Ideally, the bypass air is deflected 180° from completely back to completely forward. Because the air is instead deflected forward at a 45°, and because the engine core continues to create forward thrust, the thrust reversers create significantly less reverse thrust than normal forward thrust.

The thrust varies with fan RPM (N1), and ground speed and will increase if the pilot increases the engine RPM. The reverse thrust for a single engine of the Boeing 747-400,[7] a newer version of the 747 (with slightly more powerful engines), is shown in Figure 2.10. Reverse thrust, not used in the Kalitta Air accident, could have helped slow the plane. Boeing estimates that using two thrust reversers to slow the plane (without spoilers) would have shortened stopping on a wet runway from 152 knots by 553 feet.

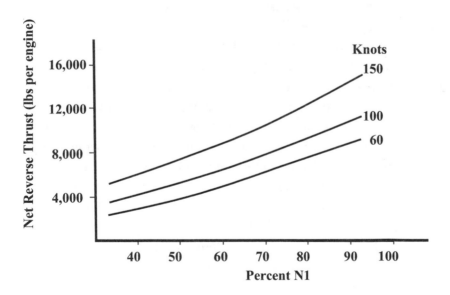

Figure 2.10. Reverse thrust of Boeing 747-400 engines. Reverse thrust force varies with plane speed and fan rotations per minute (N1).

Wheel Brakes

The wheel brakes provide most of the stopping force—over 70%. Disc brakes in a car consist of one rotor, keyed to rotate with the wheel, and one stationary brake pad (called stators on a plane). That's not enough braking force to stop a big plane. Each wheel on a big jet has a stack of rotors and stators. The Boeing 787, for example, has five rotors and four stators in each wheel. One reason big jets have so many wheels is to provide more braking capacity. Compared to the 787, the Airbus A380, the largest and heaviest commercial jet, is 233% heavier with 250% more wheels (not counting the smaller nose gear).

Pistons pressurized with 3,000 psi hydraulic fluid press the stack of rotors against the stators, creating massive frictional forces between the rubbing surfaces. All this concentrated frictional energy comes at a price—heat generation that can create problems and even ignite fires.

The conservation of energy illustrates the braking problem. The kinetic energy of the plane's speed is converted, via friction, into heat energy in the brakes. Before writing the conservation of energy equation, specific heat needs to be defined. Specific heat is the heat energy per unit mass required to raise the temperature one degree.

The plane's kinetic energy can be described by the equation ½ mass of plane × velocity^2. This energy must be dissipated by the brakes, which heat up according to the conservation of energy equation kinetic energy = specific heat (per unit mass) × mass of the heat sink × temperature rise of the heat sink. The maximum certified brake energy defines a maximum speed from which the plane can safely stop. This speed must be greater than V_1. Braking capacity can limit takeoff during hot ambient conditions and on long runways (more tire flex creates more heat energy). If braking capacity limits, V_1 must be reduced, which is normally achieved by reducing weight (normally, cargo gets removed first).

For many decades, the standard brake (and clutch) material for cars and planes was proprietary sintered steel/ceramic material. At extreme temperatures, brake fade will occur with this material. During brake fade, the surface begins to melt. A film of molten material acts as a lubricant between the stators and rotors and significantly reduces the braking force. Excess heat can even weld the brakes together to the extent that the plane cannot be moved. As described on page 65, brake fires during a rejected takeoff have been known to happen. Brake grease, tires, and any leaking hydraulic fluid or fuel will burn.

Military planes began using carbon brakes in the 1970s. Carbon brakes, which are significantly more expensive, first appeared on wide-body commercial planes in the 1980s. The major advantage of carbon is weight reduction. The Airbus A300-600, with a maximum takeoff weight of 378,500 lbs (its first commercial flight was in 1983, and carbon brakes were first used in 1985), saved 1,180 lbs in the brakes and an additional 53 lbs in the wheels. Other advantages include longer life, the brakes will not weld, and the aircraft is significantly less susceptible to brake fade. The wheels and carbon brakes for a McDonnell Douglas MD-11 are shown in Figure 2.11.

Figure 2.11. Wheel and carbon brake stack for a McDonnell Douglas MD-11. An MD-11 is essentially an updated DC-10. Wikimedia/Donzey

Carbon brakes save weight by operating at a higher temperature—2,500°F for carbon versus 2,000°F for steel. Because carbon brakes use the same hydraulics, wheels, and tires, they offer no additional braking capacity—the design braking energy is the same. Smaller planes had a harder time justifying the increased cost of carbon brakes. The Boeing 737, for example, did not use carbon brakes until 2008. The 700 lb weight reduction results in a 0.5% fuel savings. All the stories in this chapter involve older steel brakes, which are still common on many planes.

Tire-Pavement Interface

The maximum braking force occurs when the tires slip 10%–15%. (One hundred percent slip is locked brakes, where the tire slides without rotation.) A Boeing 747 tire with a 49-inch diameter rotates at 686 RPM when the plane travels forward at 100 mph. Ten percent slip means the tires are turning 10% slower (617 RPM) because of the wheel brakes. The interaction at the tire-pavement interface is more complicated than a simple sliding motion and depends on the forces of adhesion and hysteresis. Adhesion is the molecular bonding that occurs between the rubber and pavement that alternately stretches and relaxes the rubber molecules. Hysteresis is the energy dissipated by the alternate stretching and compressing of the rubber as it enters and leaves the contact patch. Hysteresis can be demonstrated by loading a rubber pad. A plot of the load versus displacement (showing hysteresis) is shown in Figure 2.12. (Note: The area under a load versus displacement plot measures the energy contained in the structure.) The loading force exceeds the unloading force because of internal friction in the rubber. You don't get back what you put in.

The braking torque from the wheel brakes opposes the tire's motion. The rubber ahead of the contact patch is stretched, and the rubber behind the contact patch is compressed. Hysteresis makes the input compressive force greater than the output stretching force, creating a net force at the pavement interface. All of this is affected by the temperature of the rubber, which is in turn affected by heat generation and heat dissipation within the rubber.

Adhesion is disrupted by a wet surface. Boeing estimates the braking

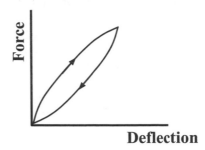

Figure 2.12. Demonstration of hysteresis when loading a rubber pad.

forces are reduced from 28% of the plane's weight to 20% on a wet runway. This is not to be confused with hydroplaning. Hydroplaning occurs when the tire floats on a film of water with a significant loss of braking force. The runway for the Kalitta rejected takeoff was listed as wet; the pilots said the runway looked dry. Runways will degrade with rubber deposits and must be periodically cleaned.

Kalitta Air Stopping Distances

Boeing estimated the stopping distances for the accident plane for the following combination of circumstances: V_1 on a wet runway (138 knots), V_1 on a dry runway (148 knots), and the maximum speed obtained during the rejected takeoff (152 knots) for a wet and dry runway (see Tables 2.1 and 2.2). Also considered are various combinations of using (or not) spoilers and two thrust reversers.

Consistent with our estimates that the braking forces for the spoilers and thrust reversers are small compared to the wheel braking forces, there is only a modest reduction of the stopping distance when using the spoilers plus two thrust reversers versus not using either. The stopping distance is only increased by up to 4% on the dry runway and up to 11% on the wet runway when not using all the deceleration devices.

The prediction of stopping distance is not precise for a variety of reasons. The braking force is affected by the tire-surface interface and is sensitive to changes in micro- and macro-surface texture. Wind gusts come and go, and pilots don't steer perfectly straight.

Table 2.1. Boeing Estimate of Stopping Distances: Dry Runway

RTO Speed (knots)	Spoilers Extended		No Spoilers	
	Two Thrust Reversers (feet)	No Thrust Reversers (feet)	Two Thrust Reversers (feet)	No Thrust Reversers (feet)
138	7,359	7,546	7,444	7,650
148	8,503	8,719	8,614	8,853*
152	8,988*	9,215*	9,110*	9,363*

*Greater than the length of runway from B1 intersection (8,777 feet).

Table 2.2. Boeing Estimate of Stopping Distances: Wet Runway

RTO Speed (knots)	Spoilers Extended		No Spoilers	
	Two Thrust Reversers (feet)	No Thrust Reversers (feet)	Two Thrust Reversers (feet)	No Thrust Reversers (feet)
138	8,135	8,469	8,536	8,974*
148	9,371*	9,751*	9,894†	10,411†
152	9,893†	10,292†	10,470†	11,023†

*Greater than the length of runway from B1 intersection (8,777 feet).
† Greater than the total runway length (9,800 feet).

The Federal Aviation Administration–mandated stopping distance calculation makes certain assumptions about how long it takes the pilot to understand the situation, reduce engine thrust to idle, and apply the brakes. In 1970, Qantas Airlines studied pilot response to engine failure in a simulator. There was a 2% chance of a four-second delay or longer between engine failure and first call out (one of the pilots announces the anomaly); and a 17% chance of an additional four-second or longer delay before a pilot responds. This means, according to this study, that there is a 0.02 × 0.17 = 0.0034, or 0.34% (3.4 times out of 1,000) chance that there will be an eight-second delay or longer between engine failure and pilot response. There are no known examples of such a lengthy flight crew response, but that's how statistics work.

Experts on human factors consider a rejected takeoff to be a "complex task"—the pilot doesn't know when the event is coming and must assess complex information with a variety of potential outcomes. The US Air Force studied pilot response to a master caution light on a cockpit panel, with the correct response being merely to push the master caution light button. Pilot response varied between 1 and 4.5 seconds, with the average being 2.4 seconds. The rejected takeoff decision is far more complicated. The pilot only has a second or two to understand the problem and decide what to do—far more involved than pushing a single button!

Captain Hedges Answers Questions about the Kalitta Air Accident

Why Didn't the Pilots Use the Thrust Reversers to Stop the 747?

This is a great question. One of the causes of the accident cited by the investigators was "less than maximum use of deceleration devices."[8] In most airliners, it is a natural and well-trained part of the rejected takeoff (RTO) sequence to retard the throttles to idle and apply reverse thrust, normally with a one-hand movement (some aircraft may require manually extending the ground spoilers between the two). Why the Captain did not use the reversers is unknown and was an error in retrospect. Additionally, the Captain stated that the spoilers were deployed correctly, but the spoiler handle was found retracted and so were the spoilers. The flight data analysis was not conclusive, although the preponderance of the evidence shows that the spoilers were likely deployed at least during a portion of the RTO.

There is another factor to consider in understanding RTO dynamics, especially when the runway is contaminated (wet or worse). Let's say, for example, that a 747 is rolling down the runway approaching V_1 in normal conditions. If suddenly the number 1 engine (left outer) quits, the plane will immediately start to yaw to the left, and the pilot will react by using right rudder to keep it going straight down the runway ("step on the good engine") before even processing what's going on. The other pilot would then call out that there's an engine failure, and the Captain would slam all four throttles closed and pull out the thrust reversers (spoilers and brakes would be used too if they weren't automatically working). As soon as the reversers came out, there would be a big control reversal—remember, after the engine failure, a significant amount of right rudder was used by the pilot to keep the plane aligned with the runway centerline when the number 1 engine failed. The application of reverse thrust causes the plane to yaw to the right because there is no reverse thrust available from engine 1; at this point the pilot immediately needs to make an aggressive left rudder input or the aircraft may depart the side of the runway. This is something most people have never considered, and it's something that makes aborts so hard in planes with big wing-mounted engines. As a side note, the Airbus 380 only has reversers on the two inboard engines (numbers 2 and 3).

In this case, reverser use was required by Kalitta in RTOs, and failing to deploy the thrust reversers lengthened the ground roll. Interestingly for most aircraft (including this 747), thrust reverser use is not factored into stopping performance calculations, although it is procedurally required. One general reason that reverser use is required in most aircraft during RTOs is because in most newer aircraft the process of activating reverse thrust will automatically deploy ground spoilers and/or autobrakes. At some airlines with diverse fleets, this is often a procedural step even on aircraft not so equipped, as it has few negative implications but makes it more consistent from fleet to fleet, making training easier for pilots changing from one aircraft type to another.

If Reverse Thrust Doesn't Work on One Engine, Can Pilots Balance Yaw with Asymmetric Braking during an RTO?

Although this might be technically possible, it is not taught because RTO is based on maximum braking; releasing any brake pressure to correct for yaw would invalidate all performance calculations and would definitely increase runway required. The good news is that the flight controls and nosewheel steering are more than adequate to provide directional control during an RTO. If asymmetric reverse causes directional control problems, the solution is to return to idle reverse thrust, which will essentially null out those tendencies. Individual airlines have their own policies on asymmetric reverse use.

Why Don't Pilots Deploy Maximum Flaps During an RTO?

Most airliners have multiple flap settings for takeoff; flaps are deflected much more during landing. Although deploying flaps to full down would increase drag during an RTO, it would also increase lift. The practical takeaway is that in a high-speed RTO, it's possible to balloon the aircraft off the runway, which is the last thing the pilots would want. During an RTO, the pilots want as much force on the wheels (and thus the brakes) as possible to stop the aircraft. As an aside, an RTO is already a confusing, jarring event, and adding procedural steps beyond the minimum required is counterproductive from a human factors perspective.

During an RTO, How Is Maximum Braking Applied?

All airliners have toe brakes on the rudder pedals. The tops of the pedals are depressed to activate the brakes on that side of the aircraft. These can be used differentially, but doing so is rare; during an RTO, maximum braking must be applied; otherwise, all performance calculations will be invalid. To achieve maximum manual braking, the tops of both rudder pedals must be held down as far as they will go until the aircraft is stopped. Most newer planes have autobrakes that can be used for RTO or landing. In these aircraft, the brakes are armed in the RTO mode at some point during the preflight or taxi out phases. Various aircraft work slightly differently, but in most planes, when a certain speed has been achieved (it's normally in the 80-knot range, coinciding with the entry into the high-speed abort regime), if the throttles are retarded to near idle or the thrust reversers are actuated, the autobrakes will apply maximum braking. Airliners have antiskid systems (similar to but more sophisticated than the antilock braking systems on automobiles) that meter braking to minimize skidding (see the section on "Tire-Pavement Interface" on page 60).

Brake Fire

The Lockheed L-1011 (first commercial flight in 1972) had steel brakes.[9] Brake temperature gauges—an uncommon option on planes in the 1970s—were available on the L-1011. But the airline declined the temperature gauges and elected to use the cockpit panel space for other equipment.

The L-1011 began its initial taxi on August 7, 1997, from Honolulu with 305 passengers and crew at 4:27 p.m. A warning light came on during taxiing, and the pilot returned to the gate. Maintenance began repairs at 5:22 p.m., including starting all three engines and taxiing out to the runway. An overheated controller was replaced, and passengers reboarded at 6:45 p.m. The plane began its takeoff roll at 7:35 p.m. The plane had by that point taxied to or from the runway three times in a little over three hours. The L-1011 attempted takeoff weighing 510,000 lbs (including 183,000 lbs of fuel) — the maximum rated takeoff weight for this plane.

The pilot noticed a cargo door warning light at V_1 (155 knots) and announced, "Door light, we're going." About a second or so later, the flight crew heard what they described as popping sounds followed by a loud shuddering noise. Air traffic controllers, perhaps a mile away, reported hearing a loud boom. The aircraft began to vibrate, shudder, and yaw to the left. The yaw was corrected with asymmetric brake and rudder inputs. The pilot rejected the takeoff about halfway down the 12,000-foot-long runway at V_R (165 knots). The plane reached a maximum speed of 168 knots.

As the aircraft decelerated to a stop, air traffic controllers reported seeing smoke and fire beneath the plane. Firefighters, responding to what they described as two explosions, dispatched themselves and were speeding toward the plane before it stopped.

The abort sequence lasted about 34 seconds. With full brakes and reverse thrust, the plane used up nearly the entire runway. After stopping, the Second Officer reminded the Captain that the brakes were probably very hot and suggested emergency evacuation. Meanwhile, the First Officer contacted the tower and requested fire and rescue equipment. The fire trucks arrived in under a minute.

The Captain was about to evacuate the plane when someone in the back yelled, "fire!" The Captain immediately ordered an evacuation. After completing the emergency checklist, the pilots and Flight Engineer entered the cabin to assist the evacuation. The flight crew departed after searching for stragglers. The firefighters arrived to find the left main landing gear engulfed in flames 10 to 15 feet high. Tires ignite at about 650°F. A 10- to 15-second blast of foam doused the fire. But the extremely hot brakes managed to reignite. A few small flare-ups were quickly extinguished. Foam was also applied to the right landing gear, which was glowing cherry red.

In the end, the "phantom" door light was attributed to fuselage flex that occurs during takeoff, when the aircraft's weight shifts from the landing gear to the wings and misalignment of a proximity sensor.

What Caused the Boom?

One of the tires burst during the takeoff roll. The sudden and violent release of any pressurized gas is dangerous. When this happens with a fuselage, it's called an explosive decompression, a process described at length in *Beyond the Black Box*. A popped balloon, fuselage, or tire all share the same basic physics. The energy in an L-1011 pressurized tire (50-inch outside diameter, 20-inch inside diameter) was demonstrated during an accident in 1980. A tire explosion ruptured the fuselage, causing two children to be ejected from the plane at 29,000 feet.[10]

Typically, a plane taxis for a mile or so and takes off. Depending on the layout of the airport, planes may taxi longer. The plane in this accident taxied out and back twice, a total distance of 11.3 miles, within a 2-hour period, with half that distance occurring 70 minutes prior to the rejected takeoff. The plane taxied a third time over a distance of about 1 mile during the aborted takeoff.

The nominal pressure in an L-1011 tire is 200 psi. (Typical tire pressures for large jet planes are 200–240 psi.) The specifications require that a tire withstand four times its rated pressure (at room temperature) for at least three seconds. Tire flexing that occurs with each rotation generates heat. Unlike automotive tires designed to dissipate heat and operate continuously, aircraft tires are designed to withstand heavy loads for short durations. Goodyear estimated the temperatures of the tire to be 190°F after the first taxi cycle, 250°F after the second, and 350°F to 400°F during the aborted takeoff. The bead (the edge of the tire that sits on the steel wheel flange) begins to degrade between 250°F and 280°F.

The failed tire had been retreaded three times (the Federal Aviation Administration permits a maximum of five retreads) and had 141 takeoffs since the last retread. Goodyear, which only manufactures new tires, does not recommend retreads and stated that expected tire life is 165 takeoffs.

Fuse Plugs

All commercial jet plane wheels are protected with thermal fuse plugs. Holes are drilled in the aluminum wheels and filled with plugs of metal that melt at low temperatures. The melting temperature of the fuse plug metal is unique for each design. The fuse plugs are intended to protect the tires and wheels from exploding with excess pressure created by extreme heat during a rejected takeoff. During the L-1011 rejected takeoff described before, two of the tires deflated after the fuse plugs melted at 338°F. The fuse plug on the tire that exploded was intact, further indicating a tire failure.

It's not uncommon for most or all of the tires to deflate during a high-energy rejected takeoff. For instance, in 2006, an Airbus A330 rejected a takeoff at just 116 knots in Brisbane, Australia (ambient temperature 82°F). Six of eight main landing gear tires deflated. The A330 fuse plugs melt at 403°F.

There is little clearance between the brakes and inside diameter of the wheels. With little circulating air, it is difficult to dissipate the built-up heat. Fuse plugs are expected to quickly melt during a high-energy rejected takeoff. During a lower-energy event, there can be a long delay before the fuse plugs melt. Experimentally measured brake temperatures are available for a Boeing 777 with a modest braking energy of 45 million foot-lbs.[11] Using the kinetic energy equation, a 500,000 lb Boeing 777 moving at just 76 ft/sec (or 45 knots) has 45 million foot-lbs of energy. In just 40 seconds, the 777's center stator heated up to 1,800°F. The brake temperature sensor (at a different location) maxed out at about 1,100°F ten minutes later. The fuse plugs reached their maximum temperatures of around 400°F forty-nine minutes later and had barely cooled off after ninety minutes.

The brakes are tested to determine the minimum brake energy that will melt the fuse plugs and the time lag it takes for the heat to flow through the wheels and melt the plugs. The Boeing 737-300 minimum energy is 23.73 million foot-lbs with a time lag of 38 minutes. The 737-300 flight manual states that the aircraft should not take off within 38 + 15 = 43 minutes after landing.

The process is not completely predictable. Sun, shade, wind, tire contact with the pavement, and ambient temperatures all affect how the heat flows through the wheels. During a rejected takeoff with an Airbus A330, one tire deflated and a second tire partially deflated from 215 to 65 psi. The fuse plug melted, releasing some pressure, and then froze back up as the wheel cooled.

It's easy to get heat into the wheels and hard to get it out, and heat-up is cumulative. On fast turnaround flights, it is possible to melt the fuse plugs on a subsequent takeoff. (Airbus even offers optional built-in brake cooling fans.) And it's difficult for pilots to notice a flat tire from the cockpit without tire pressure sensors. A flat tire during takeoff may have little noticeable effect, or cause the plane to pull to one side, or at high speeds create violent vibrations as the tire comes apart.

Fortunately, there are no known disasters in Boeing or Airbus planes caused by flat tires. An extreme example occurred on March 11, 2004, when an Airbus A300 operated by FedEx landed in Florida with the parking brakes accidentally engaged. All eight main landing gear tires went flat almost immediately, and 10% of the wheel diameter was abraded off. Other than tires, wheels, and brakes there was no other damage to the plane.

The crash of the supersonic Concorde on July 25, 2000, is the lone modern example of a tire failure resulting in catastrophe.[12] Tire failure can lead to a dangerous high-energy rejected takeoff, which is the subject of the next section.

Takeoffs That Should Have Been Rejected

Besides the explosive nature of pressurized gas, tires have also been known to chemically explode—a different phenomenon related to the rapid combustion of fuel vapor or other explosive compounds. Tire chemical explosions were believed to have resulted from accumulated hydrocarbon gases, either from decomposing rubber or brake grease. Such an accident occurred on March 31, 1986.

Just 14 minutes after takeoff, an overheated tire exploded in the wheel well of a Boeing 727 at 31,000 feet. The cockpit voice recorder captured the pilots discussing an explosion and subsequent cabin depressurization. Unfortunately, close proximity to fuel lines, electrical cables, and hydraulic lines makes the wheel well a less-than-desirable location for any damage. In this case, worse still, the ignition of spilt fuel caused a massive fire. The crew lost control and the plane crashed, killing all 167 on board. It was later determined that a partially stuck brake overheated the tire. The tire was also pressurized with air instead of more inert nitrogen. Today, to prevent chemical explosions, nitrogen is mandated for all commercial planes weighing more than 75,000 lbs.

A wheel well fire also occurred in a Douglas DC-8 in 1991. During takeoff, an underinflated tire went flat at around 50 knots. A rhythmic sound was recorded on the cockpit voice recorder. The number 2 tire was not rotating but was instead grinding down beyond the wheel tie bolts. Witnesses reported visible sparks and flames until the landing gear was retracted. After loss of pressurization and numerous alarms in the cockpit, the flight crew declared an emergency and turned back. The pilots remained unaware of any fire until alerted by the Flight Attendants. The fire consumed the fuselage floor. On final approach at an altitude of about 2,200 feet, the first of several casualties fell through the floor. Despite structural damage, the plane remained controllable until it crashed just 9,400 feet short of the runway.

In both of these examples, the fuse plugs in the tires did not work as designed. Both the DC-8 (first flight in 1958) and previous Boeing 727 (first flight in 1963) are older planes. Modern planes have brake temperature sensors, tire pressure sensors, and a thermal sensing loop in the wheel wells (and engines, too). (Temperature sensors measure at a specific location. Equipment packed into the wheel well is quite crowded, so adjacent components could easily mask a nearby fire.) An example of a thermal sensing loop is tubing filled with temperature-sensitive salt that changes electrical resistance when heated. The tubing is then snaked through a wider area wherever it will fit.

If a wheel well fire occurs, Captain Hedges advises the procedure in most aircraft is to lower the landing gear and let the fire burn itself out. The limited amount of consumable material (i.e., rubber and hydraulic fluid) will hopefully limit the size of the fire, or perhaps the wind blast can blow it out. There are no known wheel well disasters with newer planes.

Worst Rejected Takeoff

The worst rejected takeoff in the modern jet era occurred in Spain on September 13, 1982. The McDonnell Douglas DC-10-30, weighing 510,000 lbs, had every seat filled, for a total of 394 passengers and crew. The takeoff speeds were V_1, V_R, and V_2, equal to 162, 169, and 182 knots.

The pilots reported a strong vibration at or near V_1, which greatly increased during rotation. Thinking the plane uncontrollable, the Captain rejected the takeoff at 175 knots with just 4,250 feet of remaining runway (total length of 10,500 feet). The plane continued to accelerate and reached a maximum speed of 184.75 knots. The plane departed the runway at 110 knots and crashed into a concrete shed, which contained the equipment for the instrument landing system. The shed was destroyed by impact with number 3 engine. The plane continued across a highway, damaging a truck and two cars, and finally crashed to a stop (on its right side) into a concrete structure protecting a water pump on a farm about 1,475 feet past the end of the runway. Air traffic controllers witnessed the crash and notified the fire department.

Surviving an airplane crash requires a livable volume, that is, a situation where the passengers and crew are not crushed. In this case, the impact, unlike the subsequent fire, was not life threatening. In an all too common occurrence, evacuation was slowed by passengers stopping to pick up luggage in the overhead bins. A passenger sitting across from a Flight Attendant recalled, "I saw the horror on her face as she looked at the back of the plane. When I turned, I saw the smoke and flames—at first outside and then almost immediately in the cabin."[13] The fire began before the plane came to a stop and engulfed the entire plane except for the left wing. Fifty people died in the fire, eight directly; the rest were overcome by carbon monoxide. The fire is believed to have started by engine impact or after the left wing broke, spilling fuel. The plane had a full fuel load for a transatlantic flight.

The DC-10 has eight emergency exits. Exits 1L, 1R (the front two), and 2L were opened immediately. Exit 2R was later opened by a passenger, but only three or four passengers exited through it. Most of the fatalities were in the back, where the fire, smoke, visibility, and fumes were problematic. Firemen arrived in about five minutes and removed five injured passengers, some unconscious, from the wreckage.

Source of Vibrations

The right nose gear tire tread began coming apart before V_1. The disintegrating tire, now a rotating unbalanced weight, caused the vibrations. Surprisingly, the vibrations increased as the nose gear lifted off, because ground contact damps the vibrations. Loss of ground contact during rotation increased the shaking. The proximity of the flight crew to the nose gear made the vibrations feel even more severe.

The rated burst speed of the 40-inch-diameter nose gear tire (the main gear tires have a 52-inch diameter) was 235 mph (not knots). The right tire, with 14 takeoffs since its last retread, had been retreaded three times. The left nose tire had four retreads. Forensic analysis of the failed tire found defects in the retread process.

It's impossible to study rejected takeoff without concluding that a partial solution is more takeoff and less rejection. (In one study, over 50% of the rejected takeoffs should have taken off.) But it feels like second-guessing a soldier's response in combat. Professional crash investigators often reach a similar conclusion, as happened in this case. In hindsight, the DC-10 take-off should have been continued. The Spanish investigators concluded that the aborted takeoff was reasonable given the abnormal circumstances, the short time available to make a decision, and the lack of training and proce-dures on wheel failure.

DC-10 Accident Alters Rules

In 1988, a McDonnell Douglas DC-10 overran the Dallas / Fort Worth Interna-tional Airport runway during a rejected takeoff. The takeoff warning horn sounded, and the "FLAP/SLAT DISAGREE" warning light illuminated as the plane reached V_1 speed of 172 knots. The Captain rejected the takeoff. If the flaps or slats are not the same on both sides (i.e., they are asymmetrical), the plane is severely unbalanced and difficult to maneuver at the low speeds of takeoff. This is an example of an aircraft being unsafe to fly, and the takeoff should be rejected. Captain Hedges adds: It is unlikely that the slats or flaps would suddenly become asymmetrical, but this crew worked for an airline that had lost a DC-10 in a spectacular accident in Chicago in 1979 in which slat asymmetry was a key contributor. The Captain therefore chose what he thought to be the safest and most conservative course of action.

The plane decelerated normally for about five or six seconds down to a speed of about 130 knots. At that point, deceleration rapidly decreased. The plane departed the 11,388-foot-long runway at a ground speed of about 95 knots and continued across a 288-foot asphalt overrun area. The nosewheel collapsed in soft dirt and plowed to a stop about 1,100 feet past the end of the runway. Of the 254 passengers and crew, 2 were seriously injured. The plane was damaged beyond economic repair. It was later discovered that the slats were in their correct position. An out-of-tolerance position sensor triggered an incorrect warning.

The DC-10-30/40 has 10 main landing gear wheels with steel brakes, as shown in Figure 2.13. Because of lined-up pairs of tires, only six skids marks (indicating maximum braking) were found on the runway, beginning at about 3,500 feet from the runway's end. After about 1,500 feet, the skid marks began fading. Except for one set of tires, the skid marks faded to a minimum or none at all about 1,000 feet from the runway's end.

The accident flight weighed 557,900 lbs and reached a maximum speed of 178 knots, equal to a kinetic energy of 775 million foot-lbs, or about 86% of the certified braking capacity. The disappearing skid marks represented total brake failure in eight of the ten main landing gear brakes after just 36% of the design braking energy was dissipated. The brakes did not stop the plane. The plane was stopped by the nose gear plowing up soft dirt.

Maximum brake energy certification testing was completed with new brakes. Worn brakes have up to 30% less mass and therefore heat to a high-

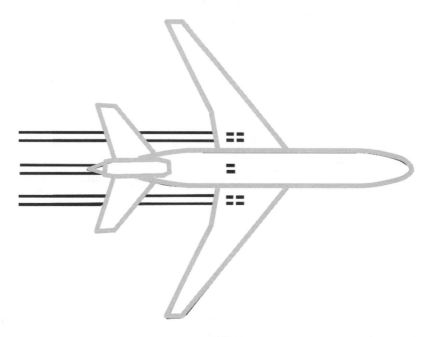

Figure 2.13. The 10 main landing gear wheels on the DC-10-30/40 (and the six runway trend marks) during the 1988 rejected takeoff in Texas.

er temperature for the same energy input. Higher temperatures result in higher wear rates.

The DC-10 brakes consisted of a stack of five steel rotors and four steel stators. Because of greater than expected brake wear during the rejected takeoff, the pistons compressing the stack of rotors and stators extended beyond its designed travel. The O-ring seals escaped, allowing hydraulic fluid to leak. Complete brake failure resulted.

The investigators reviewed the brake manufacturer's rationale for wear limits. The design engineers said their concern was potential damage and costly repairs if the friction material wore away completely. Brake performance and brake wear with less friction material were not considered by the original designers or by the Federal Aviation Administration.

The airline reported that normal brake life was about 1,000 landings. The eight faulty brakes had between 762 and 1,043 landings. But a normal landing (with less weight after fuel burn) is less severe on the brakes than a rejected takeoff.

Surprisingly, this is the only rejected takeoff accident known to be caused by brake failure. Nevertheless, after the 1988 accident, brake lining limits were changed. Certification testing of new plane designs required rejected takeoff testing with worn brakes. Existing planes were eventually retrofitted with brakes that met their rated capacity when worn.

Maximum Brake Energy Certification Test

The brake energy certification test requires stopping at the maximum possible kinetic energy that can occur during a rejected takeoff. The certification test now requires using brakes worn to their legal limit. Before the test, the brakes are warmed with "normal use," defined to be a 3-mile taxi including three taxi stops. Testing is also done with the center of gravity (CG) at its forward limit. A forward CG makes stopping more difficult because less weight is on the main gear.

A small brake fire is allowed during the Federal Aviation Administration–mandated brake test provided the fire is confined to the wheels/brakes/tires and does not spread within five minutes, which is the maximum anticipated arrival time for the fire trucks. The tires or wheels cannot explode during the certification test.

In 2011, the new Boeing 747-8 completed its rejected takeoff certification test. The plane was loaded to its maximum takeoff weight of 975,000 lbs. At 173 knots and full throttle, the test pilot slammed on the brakes (machined down to their legal limit) and stopped the plane without thrust reversers. The carbon brakes glowed red at a temperature estimated to be greater than 2,500°F. Firemen were on standby, waiting the mandated five minutes before applying water to cool the overheated equipment. The tires and brakes were damaged as expected; the rest of the plane remained intact. The fuse plugs activated as planned.

As with everything else associated with commercial aviation, additional sensors have increased safety. Brake temperature and tire pressure sensors went from nonexistent to optional to standard equipment with each generation of plane.

Captain Hedges Explains Rejected Takeoffs: To Go or Not to Go, That Is the Question

When people think about time-critical situations faced by airline pilots, they typically think of things like engine failures, cabin decompressions, and fires. Those are all serious emergencies that require prompt attention, but a more insidious threat that is more likely to be faced by pilots is the decision to discontinue (or abort) a takeoff once it has been started because of an abnormal event. Pilots routinely train for these events in the simulator, and they are among the most challenging decisions routinely trained for. They are certainly among the most time-sensitive decisions. When the *FAA Takeoff Safety Training Aid* was updated in 2004, there were approximately 18 million airliner takeoffs and around 6,000 rejected takeoffs (RTOs) annually. Precise numbers are impossible to know because low-speed RTOs are rarely reported and so are absent in the statistics. Two things are clear, however: first, based on those numbers, about 1 out of every 3,000 takeoffs ends in an RTO, and second, an increasing number of takeoffs will lead to a larger number of RTOs if the RTO rate stays the same. The type of flying a pilot does dramatically alters their exposure to this threat: a pilot flying a B-737, MD-80, or regional jet could perform 80 takeoffs in a month (though most will typically perform fewer), while an international pilot in a plane like a B-777, B-747, or A380 might perform only 8 takeoffs in a month (and conceivably fewer). This means that an international pilot might only see a real RTO situation once in a career, while a domestic pilot might be faced with the RTO decision every three or four years. That these numbers are not precise is of no real consequence; the trend is clear that pilots flying multiple flights a day will at some point almost certainly be challenged by an RTO decision. The good news is that between the 1960s and 1990s there has been a 78% decrease in the rate of RTO incidents and accidents, primarily owing to a better understanding of the high-speed performance issues inherent in jet operations, industrywide analysis, collaboration on best practices, and improved training for pilots.[14]

Regardless of the type of flying, every pilot on every takeoff must be mentally prepared to abort the takeoff if necessary, and must have a plan of action in place before advancing the power on the runway. Airlines now have detailed policies on RTO execution, and most airlines have adopted a policy that only the Captain can call for an RTO and execute it: if the First Officer (FO) were flying the aircraft prior to the abort, the Captain would assume control with a definite statement of intent (e.g., "Abort. I have the aircraft" or something similar, depending on the airline's specific policy). This policy of the Captain being the only one authorized to decide to abort the takeoff was implemented primarily because of the small time window available to make the decision, and the need to have one definite decision maker in charge. The Captain is the obvious choice because normally the Captain is the most experienced pilot on the flight deck. Airlines now require pilots to brief the runway-specific abort plan prior to takeoff and in elaborate detail on the first flight of the trip or with a new pilot joining the crew. Considerations are things like weight, speed, inoperative equipment, weather, wind, and runway length, to name a few. Every takeoff is different, and if any specific threats are identifiable, it is good practice (and at many airlines procedure) to brief them as well. A takeoff on a short snow-covered runway has a whole different set of threats than a heavily loaded aircraft departing in the heat of summer at a high-altitude airport in mountainous terrain.

Fundamental to every takeoff are a set of speeds designed to ensure the aircraft can get safely airborne or has sufficient room to abort the takeoff, depending on where the aircraft is in the takeoff run. The most critical speed for the pilot to know is V_1. V_1 has often been described as the "decision speed," at which the pilot must decide to continue the takeoff or execute an RTO. This is close but not quite right; V_1 is the maximum speed at which the RTO maneuver can be initiated and be guaranteed to stop in the remaining field length (if the RTO is executed properly). There is a nuance: once V_1 arrives, there is no more decision time available. In other words, at V_1, the Captain must be in the process of beginning the abort; retarding the throttles to idle is a universal first step in the RTO process. Calculation of V_1 can be a complex task, with multiple considerations involved. Most airlines now use computerized field analysis for better accuracy. It is possible to get valid results with paper charts, but frequently the interpretation of those charts and the corrections to them can be difficult. Typical things that get factored in to V_1 are the flap setting used for takeoff (most airliners can use multiple settings), temperature, altitude of the airport, slope of the runway, runway surface condition (wet or icy, grooved or smooth), and any deferred maintenance items on the aircraft that can alter takeoff or stopping performance.[15]

Balanced field length (BFL) is also important. Bear in mind that for any set of conditions, the BFL defines the shortest runway that allows the takeoff to be aborted no later than V_1 or to be successfully flown to meet all certification standards with an engine failure occurring at or after V_1 in the BFL distance. In a genuinely balanced situation, the takeoff is critical, as aborting past V_1 will almost assuredly lead to a runway excursion. The inertia of a heavy airliner is huge, and the brakes must absorb a lot of energy to stop an aircraft from a high speed. Nearing V_1, an aircraft is traveling 200 to 300 feet per second and accelerating 3 to 6 knots per second. The practical upshot is that delaying the abort decision even a second or two above V_1 can result in running several hundred feet off the end of the runway. This is why it is so critical that the stopping procedure begins no later than V_1. Although it is a less common issue, if the pilot did not recognize an engine failure that occurred prior to V_1, there is no guarantee that the aircraft would successfully become airborne by the end of the runway either. The good news here is that in most cases, airliners are not taking off in genuinely balanced field length scenarios, as there is normally more runway available than the calculated BFL. BFL situations tend to occur with heavyweight aircraft on short runways, and a slippery runway can dramatically exacerbate the problem.

Also key is understanding that the entire concept of V_1 is based on the malfunction being an engine failure. In reality, in more than 75% of RTO accidents, full takeoff power was available. Cases like a tire failure can initiate alarming vibrations in the aircraft, but the best course of action in such a case is generally to continue the takeoff, as stopping data are invalid with fewer tires available. There are numerous examples of a tire failure resulting in inappropriate RTOs, including the September 13, 1982, accident in which a SPANTAX DC-10 aborted a takeoff above V_1, resulting in a runway overrun, collision with a building, a fire that destroyed the aircraft, and the deaths of 50 people. Although the RTO decision was in retrospect incorrect, the investigating commission found that the decision was reasonable based on the irregular circumstances the crew faced, particularly that the crew had not been trained to deal with tire or wheel failures. To emphasize the point about the importance of making the abort decision prior to V_1, consider that this DC-10 had a V_1 of 162 knots, and though the vibrations were initially felt around V_1, the maximum speed recorded on the flight data recorder was 185 knots (remember that in this accident, like in most RTO accidents, all the engines were working correctly). Thanks to this accident and others like it, airlines now provide training on RTO scenarios that go far beyond the engine failure scenarios of years past. Pilots still practice engine failures when going to simulator training, but many more events are trained today, such as tire failures, electrical failures, fires, wind shear, instrument failures, and so on. Depending on the speed at which the malfunction occurs and some environmental factors, different decisions may be appropriate at different times. For example, on a long, dry runway, having a major electrical failure at 50 knots would likely warrant an RTO; low-speed aborts are relatively low threat and simple to accomplish. With the same failure 5 knots before V_1, a more prudent course of action would likely be to continue the takeoff. Although it may seem counterintuitive to want to fly in an airplane with a problem, at that point stopping is a statistically worse option, and there may be complicating factors that the pilot can't possibly analyze in the time available (e.g., did the electrical failure render the aircraft's antiskid brakes inoperative, resulting in far worse stopping performance?).

In recent years, aircraft manufacturers, airlines, and regulators like the Federal Aviation Administration have moved toward dividing RTOs into "high-speed" and "low-speed" regimes. Different manufacturers and airlines define the two terms slightly differently, but the consensus is that the line is somewhere between 80 and 100 knots. In newer planes, some caution and alerting systems are inhibited above 80 or 100 knots to minimize the threat of an inappropriate RTO. According to Airbus, 8% of RTOs are performed above 100 knots, and these are disproportionately more likely to result in a runway overrun or aircraft damage (brake fires are common after a high-speed RTO, and tire deflation is virtually assured). Interestingly, over 50% of runway overruns after an RTO have occurred when aborting above V_1, which is the reason that RTO training has become more varied and intensive for airline pilots. Many airlines have policies to abort the takeoff in the low-speed regime for most any caution or warning presented, but now teach pilots that as they get closer to V_1 and enter the high-speed regime to abort only for the most critical emergencies like an engine failure, fire warning, or malfunction that leads the Captain believe the aircraft is unable to fly safely. Certainly, the SPANTAX Captain believed that the aircraft was unable to fly safely, but he had never been taught how to deal with tire failures during takeoff. This is why continuing realistic training is so important.

How Do Pilots Abort a Takeoff?

All airplanes have approved procedures for aborting a takeoff. In general terms, the problem is one of turning a machine that's accelerating and preparing to fly with wings optimized for maximum lift into a machine with as much drag and deceleration as possible, as quickly as possible. Procedurally, the Captain will announce the abort and take control from the FO if necessary. The Captain's hand is always on the throttles until V_1 (and it is removed thereafter to emphasize that the aircraft is now committed to fly), and in an RTO the throttles will be quickly retarded to idle and reverse thrust will be selected. Most aircraft have automatic systems that deploy the spoilers on the wings to increase drag, decrease lift, and put more weight on the wheels to give the brakes the ability to better stop the aircraft. Many aircraft also have autobrakes that are activated to maximum deceleration. If the aircraft isn't so equipped, the Captain will use maximum manual braking and deploy the spoilers. Once the RTO procedure is correctly commenced, the FO will notify the air traffic control tower of the RTO (in low visibility, the controllers may not be able to see the aborting aircraft and could potentially clear another aircraft to use the runway if that call isn't made). As the aircraft comes to a stop, an announcement will be made for the passengers to remain seated until the crew can analyze the situation and take appropriate action. Pilots must be cautious about possible passenger-initiated evacuations, so a prompt announcement is crucial. If a passenger opens an exit without being commanded by a crewmember, bad things can happen. They may find themselves on the wings near flight control surfaces that may move and could certainly cause severe injury to someone unaware of the danger (they are pressurized with 3,000 psi of hydraulic pressure). Even more threatening is the specter of passengers getting close to a running jet engine or brake fire. After the malfunction is addressed, the pilots will decide whether to evacuate the aircraft on the runway or to taxi (or be towed) back to the gate. In all likelihood, the fire department will inspect the aircraft and follow it back to the gate if it is not evacuated. A lot of coordination goes on between the pilots, the Flight Attendants, and passengers in a situation like this, and no two situations are the same.

Most commercial jets have thrust reversers, and they should normally be used during an RTO. Most aircraft do not use reverse thrust as part of the certified stopping distances for RTO or landing for several reasons. First, if an engine fails during takeoff, the reverser on that engine will not provide any reverse thrust during the RTO; second, pilot technique and individual aircraft characteristics may limit the amount of reverse availability. For instance, on the MD-80 series, because the engines are mounted on the aft fuselage, use of reverse thrust at high power settings (above 1.3 engine pressure ratio) will aerodynamically blank the rudder by eliminating airflow over it. This is crucial because, especially with the loss of an engine, rudder is key to directional control on the runway (a slippery runway exacerbates this effect). Whether it be for an RTO or during a landing, reverse thrust is most effective at high speed and becomes relatively trivial (especially compared to the brakes) at slower speeds.

Kalitta Air: The Song Remains the Same

The May 25, 2008, RTO accident of a B-747 in Brussels, Belgium, is a song with a classic refrain: the aircraft had a malfunction in the high-speed regime on a runway with limited length for the aircraft performance. In this instance, V_1 was 138 knots, but the abort was initiated 12 knots above V_1, a classic setup for a runway overrun (see the discussion of balanced field length above). In this case, the number 3 engine (the inner engine on the right wing) ingested a bird during the takeoff roll, causing the engine to experience a compressor stall that resulted in a loud bang due to the instantly disrupted airflow in the engine. Compressor stalls are frequently caused by ingestion of a bird, ice, or foreign object; they can also be caused by engine fuel/air mixture issues or issues inherent in specific aircraft designs (e.g., the B-727 number 2 engine is somewhat prone to compressor stalls at low speeds with a significant crosswind owing to the extensive inlet ducting to the engine). After a compressor stall, the engine will usually self-correct without any further issues. The stall itself may prove to be alarming (especially to passengers), as they are characterized by one or more loud bangs, a momentary loss of thrust that may yaw the aircraft, fluctuating engine instruments, and flames coming from either the front or back of the engine. Although alarming, these are normally not airworthiness issues and would not typically be a reason to abort a takeoff close to V_1 on a runway of performance limited length; a compressor stall in isolation is certainly not a reason to abort a takeoff above V_1. (An exception might be multiple engines experiencing stalls after ingesting a flock of birds.)

As is generally the case, there was not a single factor responsible for this accident. In the case of Kalitta, the airline had experienced a relatively high number of engine failures and shutdowns on their 747 fleet. The Captain of the accident flight had been flying the same 747 (N704CK) in Inchon and had experienced a loud bang, flash of light, and aircraft yaw. In that case, the engine did fail. Because the Captain perceived the sound in Brussels to be even louder than the Inchon incident, it may have biased him into believing that the situation was even more dire than his previous failure on the aircraft. These types of personal biases are difficult to anticipate, as everyone brings different experience to the flight deck, and previous lessons learned may or may not apply to different situations. Again, that's why realistic simulator training with varied scenarios is crucial to flight safety. As it turns out, the number 3 engine was recovering without crew intervention in Brussels, and would have been no impediment to safe flight.

Close Encounters with Rejected Takeoffs

If a pilot spends a career on the flight deck, he or she will likely encounter a rejected takeoff (RTO) decision at some point. Lots of factors figure into the outcome of the event: what was the malfunction or issue that prompted the decision? Were there multiple problems? How long was the runway? Was it wet or snowy? How heavy was the aircraft? The biggest question, though, is how well was this Captain trained and prepared for the decision?

Captain Hedges is an experienced airline pilot who has been involved in three RTOs in operational airline flying, two as a First Officer and one as a Captain. The lesson he has taken away from these three events (which span 27 years from first to most recent) is that no matter how well prepared a crew is, an RTO is always going to be a shock. It is much more surprising than when practiced in the simulator (obviously, pilots expect malfunctions in the simulator).

The first RTO occurred when Captain Hedges was a new First Officer (FO) on the McDonnell Douglas DC-9. First, some background: it was common practice on the DC-9 to apply slight down elevator (where the control wheel is pushed forward) during the takeoff roll to prevent the nose landing gear strut from repeatedly compressing and extending back and forth, which could have the effect of changing the modes of some systems from the flight to the air mode in rapid succession. It also maximized nosewheel contact with the ground, which was emphasized to new pilots on that aircraft because there is a small amount of nosewheel steering available from the rudder pedals to enhance directional control during the takeoff and landing rolls, as the rudder is fairly ineffective below approximately 60 knots. Most airliners have an interconnect between the rudder pedals and the nosewheel steering for this reason, but unlike most airliners, the DC-9 normally has an unpowered elevator. The one exception was that if a pilot pushed the elevator far forward to recover from a stall, hydraulic pressure would be applied to provide full nose-down control authority to allow recovery. When this occurred, a blue "Elevator Power On" light illuminated on the annunciator panel.

On the takeoff in question, the FO was flying the aircraft. The runway was wet and the weather was windy. Being conscious to keep the nose on the ground, the "Elevator Power On" light momentarily illuminated as the FO was applying down elevator input; the Captain did not analyze the situation correctly and called for an abort, which was successfully accomplished, although a brake and tire change was necessitated (the fuse plugs released the tire pressure safely). In debriefing the event, a few things became apparent. Unsurprisingly, a training issue was most prevalent: although the FO had been properly trained to apply nose-down elevator during the takeoff run, he had never been told how much down elevator to apply. As it turned out, he was using more than was required, which momentarily triggered the alerting light. The Captain's concern was that there was a flight control malfunction that would not allow the aircraft to become airborne owing to full nose-down elevator application, and he decided to execute the RTO, which he believed to be the safest course of action (it's better to depart the end of the runway at 50 knots than 200 knots!). If the Captain had been better trained, he might have thought to look at how far forward the FO was holding the control yoke before making this decision. This event was summarized to all DC-9 pilots, and a bulletin was issued detailing the issue. Finally, the training syllabus was changed to emphasize holding forward elevator early in the takeoff roll, but decreasing the amount with increasing speed; the proper amount of deflection was also subsequently stressed. This is the only RTO that Captain Hedges has experienced that began in the high-speed regime.

The second event occurred many years later when Captain Hedges was an FO on a Boeing 757. It was the Captain's turn to fly, and the FO was the Pilot Monitoring (PM, formerly known as the Pilot Not Flying, or PNF). On a clear day on a long runway, the aircraft was cleared for takeoff on a short flight. The Captain advanced the thrust levers and called for the autothrottles to be engaged. This would normally drive both engines to the target power setting for takeoff. At about 60 knots it became apparent from the aircraft yawing to the right and engine instrument anomalies that the right engine was not developing takeoff thrust. The aircraft was extremely lightweight; the B-757 is well known for brisk acceleration even when much heavier. Because of the light weight of the aircraft and the fact that both engines were developing power (although the right was not producing as much as normal), even with the abort begun around 60 knots, the aircraft got above 80 knots during the RTO process, and that was with prompt action taken by the Captain. This event didn't even result in overheated brakes, though maintenance on the right engine was required before the flight was able to finally depart. Both pilots discussed the abort and were both surprised how fast the plane got during the procedure. This was a key moment of clarity for that crew in fully grasping the V_1 concept: the abort must be started prior to the V_1 speed or it will be exceeded (probably by more than the pilots expect) during the RTO process, making a runway overrun much more likely.

The final RTO that Captain Hedges was involved with was his only one as a Captain. Departing in an MD-80 in good weather on a long runway, immediately after takeoff power had been achieved, the entire left electrical system failed. The failure occurred at around 60 knots, and an RTO was executed because the aircraft was clearly in the low-speed regime, and the failure was significant. This was in line with best practices for major malfunctions in the low-speed regime, and the RTO and subsequent coordination with the Flight Attendants and passengers was smooth. The aircraft was taxied back to the gate, passengers were deplaned, and maintenance action was undertaken to repair the electrical system. The flight ultimately departed the next day. Again, this was to be a short flight, and despite prompt action to initiate the RTO, acceleration was rapid (both engines were producing takeoff thrust) and the aircraft got up to 74 knots at a peak. This was further reinforcement to both pilots of the importance of the low-versus high-speed flight regimes, and the importance of not aborting above V_1. If this electrical failure had occurred above the 80-knot limit defining the high-speed regime for this fleet of MD-80 aircraft, the takeoff would have been continued and the problem sorted out airborne.

3 Controlling the Plane

Bank (or roll) is rotation about the longitudinal axis, the axis passing along the centerline of the fuselage. As the plane banks, the vertical component of lift decreases, as shown in Figure 3.1. With excess roll, the plane may lack sufficient lift to maintain safe flight. Just such an accident occurred in 2007 in Cameroon. This accident occurred during climb—the phase of flight following takeoff.

Kenya Airways Flight 507 originated in Abidjan, Ivory Coast, bound for Nairobi, Kenya, with an intermediate stop in Douala, Cameroon. The 737-800, delayed by thunderstorms in Cameroon, eventually took off shortly

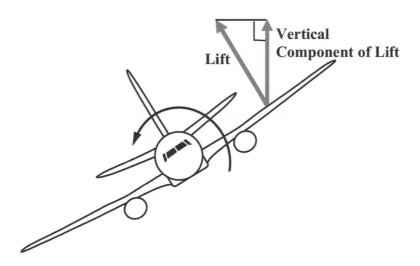

Figure 3.1. Rolling reduces the vertical component of the lift force supporting the plane's weight.

after midnight on May 5, 2007. Soon after takeoff, air traffic control lost contact with the plane.

On that stormy night, the villagers of Lolodorf described a loud boom and a flash of light. Based on these reports, the *New York Times* stated that the still-missing plane went down near Lolodorf, 155 miles south of the Douala Airport. The *Washington Post* located Lolodorf 100 miles from Douala and reported that controllers lost contact with the plane 11-13 minutes after takeoff. The *London Telegraph* said the plane sent a distress signal 40 miles from Douala. The May 8 headline in the *Los Angeles Times* read, "Engine Trouble Is the Focus of Jet Crash." Early press reports are often speculative at best and can be confusing and even contradictory.

About 40 hours after takeoff, the plane was located just 3 miles from the end of the runway in a mangrove swamp. Reporters said the crash site was 40 minutes by foot from the nearest road. The fragmented plane was found at the bottom of a water-filled crater partly covered by a thick canopy of trees. The flight data recorder (FDR) was found almost immediately. Swamp water hampered recovery of the cockpit voice recorder (CVR). The 30-foot-deep crater was repeatedly pumped out after rapidly refilling. The CVR was recovered in four separate pieces (pinger, battery, motherboard, and memory module) on June 15.

The recorders were sent to the Transportation Safety Board of Canada (the Canadian equivalent of the US National Transportation Safety Board) and analyzed in the presence of investigators from the NTSB, Federal Aviation Administration (FAA), Boeing, Kenya Airways, and Cameroon. The NTSB is the representative of the airplane's country of origin (for US-built aircraft) and will supervise any investigation that takes place at the manufacturer's site or at any US supplier. The analysis of the two black boxes completes the story.

Investigators found that the flight crew made numerous procedural errors. The Captain did not conduct a departure briefing, nor did he have a plan to avoid nearby thunderstorms. The Captain used the wrong call sign, resulting in 15 minutes of confusion and delay with air traffic controllers. The "before taxi" checklist was called for, but the copilot instead began the "before takeoff" checklist, and the Captain did not correct the mistake. The plane took off without asking for or receiving permission from controllers.

After takeoff, the plane had a slight tendency to roll to the right. This not uncommon effect can occur from manufacturing misalignments. For the first 24 seconds of flight, the pilot easily corrected a slight roll by turning the control wheel to the left. The pilot pressed the button to engage the autopilot, but the autopilot did not engage. Apparently unnoticed by the flight crew, the plane by that point had banked itself 11° to the right. For the next 55 seconds, the plane was flown neither manually or by autopilot.

Standard procedure calls for the Pilot Flying to engage the autopilot and call out the action so that the other pilot can verify that it occurs—a check and cross-check for this critical function. Neither pilot noticed that the autopilot had not correctly engaged.

Why didn't the autopilot engage? Two theories were explored: either the pilot simultaneously nudged the control column (the autopilot gives preference to human commands), or a failure occurred in the autopilot system (between December 2006 and February 27, 2007, there were several occurrences of autopilot malfunctions in this aircraft). But the autopilot had passed two routine inspections before the accident.

The working theory was that, while trying to maneuver through storms, the flight crew had their heads buried in the weather radar screen. The plane continued its roll and was simultaneously turning. (The turn rate increases with increased bank angle.) Just 24 seconds into the pilotless event, the bank angle was 24° to the right. The heading selector on the autopilot had just been changed by the Captain from 120° to 165°, but because the plane was turning itself, the actual heading at this point was 190°. From the crash report: "The behavior of the flight crew during these [pilotless] 55 seconds demonstrate a lack of rigor in piloting, . . . confusion in the use of the autopilot, and poor situation awareness."[1] The flight crew should have noticed the plane was banked 24°, headed in the wrong direction, and flying without the autopilot.

Eventually, the bank alarm (a recorded message with the actual spoken words "bank angle") sounded as the plane banked 35° to the right. The clearly startled pilot performed a series of left–right wheel inputs, which were mostly to the right and actually worsened the bank angle. Just 3 seconds later (1 minute and 19 seconds after takeoff), the bank angle was 55° and the altitude was 2,700 feet. With a lack of external references (nighttime flight over an uninhabited area), the Captain was spatially disoriented. Human factors engineers estimate that a spatially disoriented pilot needs 10-35 seconds to recover. When the pilot finally looked up, the plane was not doing what he expected, and worse still he didn't understand why. The plane was then 23 seconds from impact and seconds from an unrecoverable situation.

The Captain successfully engaged the autopilot and "resumed his confused movements of the flight controls,"[2] including applying hard right rudder. Computer simulation by Boeing concludes a hard and sustain right rudder pedal input when the plane is banked 50° will cause the nose to drop abruptly. The plane eventually reached an altitude of 2,900 feet with an 80° bank angle before it began a spiral dive into the ground. Impact occurred just 1 minute and 42 seconds after takeoff with a vertical sink rate of 14,000 feet per minute (138 knots vertically), airspeed of 287 knots, and a nose-down pitch of –48°. The impact speed and orientation greatly exceeded any expectation of survival. The crash showed evidence of an instant fuel-sprayed fireball that was perhaps rapidly extinguished by the swamp and rain.

The 52-year-old pilot had over 8,000 hours of experience, including 823 hours in a Boeing 737. In 2002, examiners noted deficiencies in his 737 certification upgrade to Captain, including: crew resource management (CRM), knowledge of airplane system, adherence to procedures, cockpit scanning, situational awareness, and decision making. Similar problems were noted on yearly recertification examinations. Because of these deficiencies, the Captain required additional training in 2004 and 2006.

Captain Hedges Explains: The Cameroon Accident

Why Didn't the Pilots Notice the Plane Rolling?

The plane's motion added to the confusion. With the plane turning in a "coordinated turn" (a topic discussed at length later in this chapter), the G loads on the pilots would not have seemed out of the ordinary despite the uncommanded and excessive roll. In a coordinated turn, there is no detectible side loading; the feel of gravity is still acting through the floor of the aircraft, creating the illusion of level flight (i.e., there is nothing analogous to a car passenger being shoved into the door on a tight, fast turn).

Although vision is the strongest sense for detecting location in space, humans also have a sensitive vestibular system in the inner ear that is controlled by three fluid-filled semicircular canals that act as accelerometers. These are exquisitely sensitive and serve us well so long as we have visual contact with the horizon. For pilots, flying in clouds or at night is hazardous because without the ability to align visually to the horizon, numerous illusions can occur; they can happen insidiously or quickly and can be overpowering. This is called spatial disorientation. All senses have minimum thresholds to generate a neural signal. Motion below this threshold does not register with the brain, and errors tend to accumulate unnoticed. If a plane begins rolling at a relatively small rate, it is entirely possible for the pilots not to notice the sensation if they are not carefully monitoring their instruments. When something suddenly signals a disagreement between their perception and reality (in this case, the aural bank angle warning), it takes pilots a while to understand the information and take appropriate action. That's why it's vital that one of the pilots be designated to fly the aircraft and keep a good instrument cross-check at all times.

It is key to safe flying to always know your location and orientation in space; particularly in instrument conditions or at night, it is imperative that pilots continuously monitor their gyroscopic flight instruments. A good instrument scan is important for pilots, and modern aircraft like the B-737-800 have large, easy-to-interpret instruments. (See Figure 3.3 and note the large attitude director indicator at the center of the primary flight display mounted directly in front of the pilots; this instrument is the center of a good instrument scan and shows the pitch and roll of the aircraft relative to the horizon.) Instruments must be monitored continuously and discrepancies quickly addressed. If the Captain had stronger CRM and threat and error management (TEM) skills, he would have deliberately slowed the operation down, analyzed the weather while on the ground, formulated a plan, formally designated a pilot to fly, and once airborne ensured that someone was continuously flying the aircraft and monitoring the instruments.

The Kenya 507 pilots took off hastily into bad weather at night over featureless terrain, a scenario that is a recipe for spatial disorientation. Central to staying oriented is a good instrument cross-check. From the beginning of instrument flight training, it is continuously preached to pilots to trust their instruments. It is a credit to the profession that most pilots are good at this, but lack of situational awareness has been the cause of several high-profile accidents. Unfortunately, these accidents tend to involve loss of control (LOC), where a pilot becomes disoriented at altitude and loses control, generally resulting in an impact with terrain that is unsurvivable. For that reason, airlines have been putting a lot of emphasis on "back to basics" instrument scan and proficiency in manually flying the aircraft without autopilot in recent years (particularly since the loss of Air France 447 and Kenya 507, along with several other accidents in the same time frame).

Do the Instruments Indicate Roll Angle?

All airliners have a prominent attitude director indicator (ADI) on each pilot's instrument panel (there is a third standby indicator that can be used as a tiebreaker in case of disagreement between the two primary ADIs). The ADI has an aircraft symbol from the pilot's perspective that is fixed; behind that symbol a movable horizon line is displayed. Above the horizon is colored blue to represent the sky, below is colored brown or black to represent the ground. The airplane's position in pitch and roll (known as attitude) relative to the horizon is interpreted by looking at the position of the aircraft symbol relative to the horizon line. This is the primary skill of instrument flying. Older ADIs were driven by mechanical gyros; modern aircraft generally have ADIs that get their data from extremely accurate laser gyros.

How Does Instrument Scan Work?

A good instrument scan is fundamental to safe flight. Pilots learn this skill based on the layout of the instruments, and there are two fundamental formats on airliners. In earlier aircraft like the 747-200 discussed in Chapter 1, there are six primary instruments, each indicating a primary parameter, located in front of the pilot in a standardized layout. The ADI is at the top center and is the primary focus of instrument flying, as it shows aircraft pitch and roll relative to the horizon. To the left of the ADI an airspeed indicator shows indicated airspeed (and in most planes Mach number as well), and to the right of the ADI an altimeter shows airplane altitude relative to sea level. Directly below the ADI is a horizontal situation indicator (HSI), which is a compass with a course indicator integrated into it so that the pilot can see at a glance where the aircraft is relative to the desired course. To the right of the HSI is a vertical velocity indicator (VVI, though it goes by several other names too), which shows rate of climb or descent, a vital parameter for instrument flying. To the left of the HSI is another compass card from a separate compass system so that pilots have a redundant source of heading information. The older-style analog 737 cockpit with these instruments is shown in Figure 3.2. This is the type of instrument panel that almost all pilots began their career with, and the instrument scan is generally a hub and spoke pattern with the ADI the hub. Pilots will look at the ADI, then quickly look at the altimeter, then back to the ADI, then to the airspeed indicator, then back to the ADI, and so on. The emphasis placed on the instruments other than the ADI may be different depending on phase of flight, but the ADI is always the primary focus of the scan. Although pilots quickly complete this scan, sweeping past various instruments in fractions of a second, there is still a considerable amount of real estate to visually cover, so engineers and human factors experts sought to make this information easier and quicker to absorb. Cross-checking other instruments and engine indicators is also done periodically in relatively rapid succession. This is a fatiguing process, so the use of the autopilot is encouraged (and sometimes mandated) after takeoff and during the climb, cruise, and descent portions of most flights. With the autopilot on, the pilot controls the action of the aircraft with the autopilot controls, but the actual manipulation of the controls is handled by the autopilot. In the Cameroon 737 accident, the Captain was unable to get the autopilot to engage in the "Command" mode, which would allow for full control through the autopilot interface.[3]

Why discuss the older format when most aircraft have now progressed to a more condensed electronic format? Simple. Both pilots of Kenya 507 (and most of the other accident pilots in this book) came from a background of flying aircraft with the conventional format. The Captain had been a longtime 737 pilot, having initially flown the earlier B-737-300 with a conventional instrument layout (a hybrid format seen on earlier B-737s and B-757/767s has an electronic ADI and a small moving map in lieu of the HSI, but the information presented and instrument scan required are similar to that of an older completely analog aircraft). The First Officer (FO) was a young and relatively inexperienced pilot with experience with the conventional format.

Both pilots had originally qualified at Kenya Airways on the B-737-300 with a conventional instrument layout. Subsequently, Kenya Airways began operating B-737-700 and B-737-800 aircraft. These aircraft had six large liquid crystal displays (LCDs) instead of the six separate instruments (see Figure 3.3), but, curiously, Kenya Airways configured the displays differently between the two new types despite the almost identical physical cockpit installations. For the B-737-700, Kenya Airways specified that the displays have a configuration commonly called EFIS/MAP, which emulated earlier aircraft with six separate instruments on the LCDs. Boeing offered this option to enable easier transition in instrument scan for pilots transitioning from earlier B-737 models, and to allow pilots to more easily fly both older and newer models in a single fleet. When Kenya Airways received their B-737-800s, they configured the displays in the newer PFD/ND format, as discussed on page 93. After training, pilots were expected to fly all three B-737 models in the fleet interchangeably despite the large differences in instrumentation and scan requirements. At the time of the accident, the Captain had logged 8,682 total flying hours, including only 158 on the B-737-800, while the FO had logged 831 hours, with 57 hours on the B-737-800. Both pilots were much more experienced with the conventional display formats and scan requirements, though how much that affected their inability to safely fly the accident aircraft can never be precisely quantified. Old habits die hard, and this crew found themselves in a situation they were not prepared for, using instrumentation with which they were relatively inexperienced. Significantly, in the Kenya Airways cockpit configuration, the autopilot command "CMD" indication is in different places on the B-737-800 than earlier models. This is a crucial point from a human factors perspective, as once the crew believed the autopilot was in Command, they effectively stopped flying the aircraft and monitoring their attitude; the aircraft had nobody flying it. The fact that the crew failed to notice that the autopilot was not engaged as desired is a crucial link in the accident chain. It was a perfectly safe, well-maintained aircraft, but with nobody minding the store it got into an unusual attitude, the crew became rapidly disoriented, and the Captain subsequently made control inputs that greatly exacerbated their situation until the aircraft was in an unrecoverable state.

Newer aircraft like the A320, B-777, and B-737-800 have an improved instrument layout consisting of six 8-inch by 8-inch LCD or CRT (cathode ray tube on older planes) screens in front of them, called the primary flight display (PFD) and the navigation display (ND), as shown in Figure 3.3. The PFD and ND are so fundamental to safe flight that they are duplicated for each pilot.

During transition training from the older display format, pilots are trained and tested on their ability to effectively fly with the PFD/ND. Most quickly embrace it, as the organization of the displays is fundamentally

Figure 3.2. First-generation Boeing 737 analog cockpit. Wikimedia / Sol Young

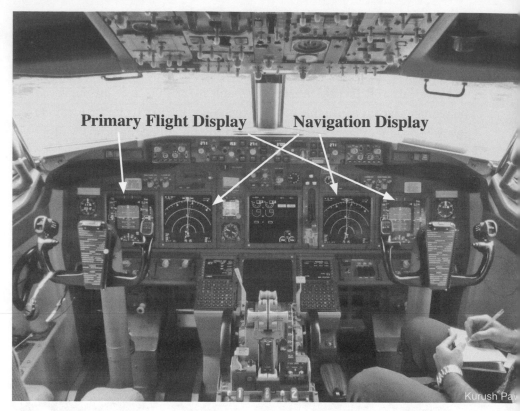

Figure 3.3. Digital cockpit in a Boeing 737-800. Wikimedia / Kurush Pawar

more compact than the original array of six instruments, the scan is quicker, and deviations are more readily observed.

The ND is a large pictorial map of where the aircraft is relative to important things (navigational aids, the selected route, airports, etc.) and hugely enhances the situational awareness of the pilot. With older instruments, it was up to the pilot to build an accurate mental model of where he or she was in relation to everything (e.g., the intended course, terrain, etc.) in time and space, but the ND greatly simplifies this task. Equally revolutionary is the PFD, which still has at its center the attitude director indicator (ADI), which is even more prominent than in older aircraft. On either side of the attitude indicator are vertical "tapes" that measure altitude, airspeed, and vertical speed (climb or descent), as well as a compass readout below the attitude indicator. When flying with the PFD, the scan is much faster because there is less physical distance between relevant readouts. In an aircraft that is in straight, unaccelerated, level flight, the attitude indicator will show the aircraft in whatever the level flight attitude is (this depends on several factors; most jet airliners are slightly nose-high during level flight), and the tapes for altitude, airspeed, and vertical speed will be centered at the desired parameters and not moving. It is easy to notice if, for instance, the aircraft begins a climb, as the altitude tape will show an increase in altitude, and the vertical speed tape indicator will show a climb; additionally, if no power is added, the airspeed will begin to decrease. Minor perturbations are therefore much more rapidly noticed and corrected. Although the shape of the instrument scan is different, the ADI is still the central instrument to a safe scan. For commonality, the airspeed tape is still on the left of the ADI, and the altimeter is on still on the right. With the PFD and ND the scan becomes straight across the PFD and back because the displays are compact. Because awareness of automation modes is so critical in advanced aircraft, manufacturers put a flight mode annunciator (FMA) at the very top of the PFD, so that pilots can know what autopilot, autothrottle, and navigation modes are in use: this is where the pilots of Kenya 507 should have verified that the Command mode was engaged. The ND provides a large map with the weather radar overlaid on it so that pilots can see real-time weather displays in relation to navigational aids, airports, and the route programmed into the flight management computer, which is displayed as a magenta line (green on Airbus aircraft) on the ND.

Regardless of the way information is presented to pilots, the key to a safe flight, particularly in instrument conditions, is to have one pilot dedicated to flying the aircraft and for that pilot to maintain a rapid instrument cross-check and good situational awareness of where the airplane is and how it is performing. This was a clear deficiency in the Kenya Airways accident.

Why Did the Cameroon 737 Tend to Roll to the Right?

Some aircraft are "bent," meaning that with the controls neutral they have a tendency to fly in an attitude that is not straight and level. This is relatively common for older aircraft and designs in particular, as manufacturing processes decades ago required shimming to get the planes to measure out exactly right. Modern technology, which uses laser levels and the like, has improved the situation, but years of operation tend to make aircraft minutely different. This is normally corrected by skilled maintenance personnel who periodically "rig" the aircraft to fly straight with no slip, skid, or roll tendencies when the controls are neutral. Manufacturers specify maximum acceptable deviations, and some aircraft come from the factory requiring small amounts of trim. All aircraft have rudder trim available to help control unwanted yaw tendencies, and most, including the 737 series, has aileron trim to help cancel out unwanted rolling tendencies. These are actions that are innate to experienced pilots, and it is common to use rudder and aileron trim in small amounts to make the plane fly "straight." This opportunity was available to the Kenya Airways pilots, but they were preoccupied and allowed the minor and readily controllable rolling tendency get out of control until it reached the point they were unable to successfully recover the aircraft.

The investigation concluded that the right-rolling tendency of Kenya 507 was from a combination of the aircraft's construction and a slight amount of right rudder trim (0.13°) present during takeoff. During takeoff and immediately thereafter, the Captain had no problem controlling this tendency; once the crew thought the autopilot was engaged and neither pilots or automation were controlling the flightpath, the right-rolling tendency continued below the crew's threshold of perception. Then the bank angle alarm brought an urgent and unexpected need to safely recover the aircraft. A good instrument scan and effective crew resource management would have prevented the airplane from getting into such a dire predicament in the first place. It is worth noting that commercial airliners routinely use up to 30° of bank in normal maneuvering; when the bank angle alarm went off at 35° of right bank, the aircraft was barely outside of the normal maneuvering envelope and could have been easily recovered with a simple left roll input of the aircraft yoke. The actual control inputs from the Captain made the situation much worse. Jet airliners require infrequent use of rudder: the Captain's large right rudder input was improper and had the net effect of yawing (and rolling) the aircraft into a steep bank and nose-low attitude, from which recovery became quickly impossible.

Crew Coordination in the Cameroon Accident

All airline pilots undergo crew resource management (CRM) training to learn how to work effectively in a team to safely complete every flight. This training began after a series of high-profile accidents in the 1970s and is now mandatory. The Captain of Kenya Airways Flight 507 had received CRM training, but he had repeatedly demonstrated having difficulties with crew coordination on various proficiency checkrides throughout his career. Although he was paternalistic to the much less experienced First Officer (FO), the Captain didn't effectively solicit input from the FO in planning and making decisions. The Captain was considered by others to be authoritative, arrogant, and overconfident, and numerous FOs did not want to fly with him.

There was not effective teamwork in the cockpit of Flight 507, and given the lack of planning prior to departure, it is unsurprising that the pilots quickly became spatially disoriented and were unable to recover the aircraft from a series of worsening unusual attitudes, which ultimately led to a loss of control of the aircraft. The distracting presence of thunderstorms ahead of the aircraft, a dark night, and no lights on the ground—combined with a poor instrument scan after takeoff and a slight right rolling tendency of that particular 737—led to a situation the Captain was unable to cope with. The unassertive FO failed to call out operational deficiencies and errors to the Captain as was required, which was a trend that had been noted in the FO's training. The crew attempted to engage the autopilot in the Command mode, in which the autopilot would control pitch and roll of the aircraft, but it did not engage; the FO should have called this to the attention of the Captain but did not. When the aircraft's automated bank angle warning sounded, the Captain was alarmed and actually increased the right bank of the aircraft while struggling to engage the autopilot. Ultimately, the FO did tell the Captain to level the wings by rolling left, but by that point the situation had deteriorated beyond the Captain's ability to recover, and in fact the Captain made a right rudder input, causing the aircraft to rapidly exceed 90° of right bank. In normal airline operations, pilots don't exceed 30° angle of bank; at 90° of bank, there is no vertical component of lift.

There are plenty of precursor accidents with authoritarian or dictatorial Captains where crew members felt intimidated and failed to speak up when they should have; this is one of the reasons CRM (and now TEM) training is so vital. Interestingly, more accidents occur with the Captain flying than the First Officer, which underscores the importance of the Pilot Monitoring's duties: when a Captain makes an error, FOs are more hesitant to speak up and correct their superior's mistakes. But when the FO is flying, because the Captain is, by definition, in command of the flight, he or she is generally much less hesitant to correct the junior pilot. Clearly in the case of Flight 507, there was not an effective team formed in the cockpit, and that element formed a key part of the accident chain that led to the crash.

Captain Hedges Explains: Controlling the Plane

Because many of the case studies in this book contain discussions of pilot inputs, it's relevant to discuss basic control concepts and nomenclature so that readers can envision what pilots are doing (or are attempting to do) in these various scenarios.

Fixed-wing aircraft behave as though they were being maneuvered around the center of gravity (CG), a point around which the three axes of the plane rotate, intersecting at 90° to each other. For each of the three axes there is a control surface, or on large jets frequently multiple control surfaces, through which the pilot makes control inputs to control how the plane moves. Every plane has strictly defined CG limits, and exceeding these limits can result in an aircraft that is difficult if not impossible to control. The CG is calculated prior to every departure, and the aircraft is loaded carefully to ensure that the CG will be maintained within a safe envelope at all times during the flight.

The nose pitches up (or down) by rotation about the lateral axis, as shown in Figure 3.4A. The pitch (or lateral) axis extends approximately through the two wingtips, and the CG and is controlled by the elevators located on the trailing edge of the horizontal stabilizers at the rear of the aircraft. The elevators are controlled by pushing or pulling on the control yoke (wheel) in front of the pilots or, in the case of some more modern Airbus airliners, the side-stick controller (SSC, much like a computer "joystick"). In either case, pushing forward makes the aircraft pitch to lower its nose, while pulling makes the plane raise its nose.

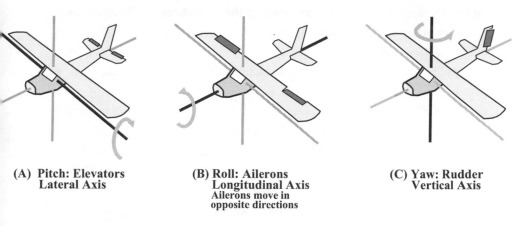

(A) Pitch: Elevators
    Lateral Axis

(B) Roll: Ailerons
    Longitudinal Axis
    Ailerons move in
    opposite directions

(C) Yaw: Rudder
    Vertical Axis

Figure 3.4. An airplane's three types of rotation, the control surface activating the motion, and the named axis of rotation.

A plane rolls or banks its wings by rotation about the longitudinal axis, as shown in Figure 3.4B. The roll (or longitudinal) axis runs from the nose, through the CG, to the tail of the aircraft. Movement around this axis is called roll and is commanded from the cockpit by the pilot turning the yoke to the left or right or moving the side-stick to the left or right. This results in the plane banking in the commanded direction. Almost all planes use ailerons as the basic roll control surfaces, and most jets also use spoilers (see Figure 3.5) as roll augmentation to achieve increased maneuverability. (There are some aircraft that don't have ailerons at all, using only spoilers for roll control, though this is relatively rare.) Ailerons are small hinged flaps at the back of the wings that move opposite each other when the pilot rolls the aircraft. When a pilot rolls the aircraft to the right, for instance, he would turn the yoke or move the side-stick to the right, which would (through a complex set of assemblies) cause the aileron on the left wing to go down and the aileron on the right wing to come up. When the left aileron goes down, it creates more lift on that wing, causing it to rise. On the right wing, it's exactly the opposite: the aileron going up destroys a bit of lift and that wing falls, and the net result is the airplane turning to the right. Banking the wings is fundamental to turning, as explained in the section "The Physics of a Coordinated Turn" on page 108. Most jet aircraft also have spoilers, which are hinged panels that pop out of the top of the wing to destroy lift and increase drag. Spoilers are useful to the pilots because they help slow the plane in flight and after landing, and help the plane descend faster when it needs to. They also work differentially to help turn the aircraft, and although the specifics of the systems vary between aircraft types, in general, when a pilot is commanding a significant roll, some of the spoilers on the wing going down will come up to help increase the rate of roll (huge airliners are much more nimble than most people would expect!).

Many large jets have two ailerons on each wing. In most cases, at higher speeds the set closest to the wingtips is deactivated to prevent overcontrolling the aircraft, though some aircraft (e.g., the L-1011) have all four operational at all times. The control surfaces on a modern jet are as shown in Figure 3.5.

The third axis is from the top of the aircraft to the bottom through the CG, referred to as the yaw (or vertical) axis (see Figure 3.4C). Movement around this axis is caused by the rudder moving left or right. The rudder is controlled by pedals under the pilot's feet; pressing the left rudder pedal causes the nose to slew to the left, while pressing the right pedal causes the nose to slew to the right. In modern jets, the rudder is the least used of the primary flight controls and is mostly used in crosswinds to help align the plane with the runway during takeoff and landing. Its other big function is helping maintain controlled flight with an engine shutdown (a task few pilots ever see in their airline ca-

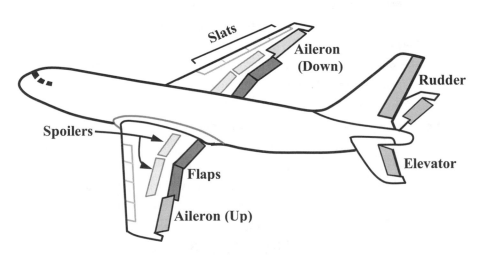

Figure 3.5. Control surfaces on a modern commercial jet.

Spoilers or Speedbrakes?

Manufacturers and airlines are sometimes inconsistent on nomenclature, and the terms "spoilers" and "speedbrakes" (or "speed brakes") constitute one of the most common areas of confusion. Although conventions vary somewhat, in this book we use these words essentially interchangeably for simplicity. They are taken to mean panels on top of the wing that are used to increase drag and/or decrease lift. Traditionally, speedbrakes are panels that can be deflected into the air to create more drag. They may be mounted on different places on a variety of aircraft: the British Aerospace BAe-146 (an uncommon British airliner), for instance, uses a pair of clamshell-shaped speedbrakes mounted at the tailcone of the aircraft. Military aircraft have a wide variety of speedbrake configurations: the Republic F-105 has four petals around the tailcone that extended at 90° angles to each other, while the F-15 has one huge speedbrake on the top of the fuselage right behind the pilot. The key similarity is that these devices all increase drag.

Spoilers are flaps that can be extended from the top of the wings that, like speedbrakes, increase drag, but they also decrease lift. Spoilers are frequently used as speedbrakes in flight, helping pilots increase their rate of descent or slow down rapidly. They are also used on the ground to help slow the aircraft on landing or during a rejected takeoff. In most designs, spoilers deploy to a greater angle on the ground than in flight (on a DC-9, for instance, the maximum spoiler deflection when used as a speedbrake in the air is 35°; on the ground it is 60°), and some panels may only be ground spoilers and not for use in flight (the MD-80, for instance, has three spoiler panels on each wing, the two farthest from the fuselage can be used either on the ground or in flight, while the spoiler nearest the fuselage is for ground use only). Spoilers can be used differentially in most airliners as well: when turning the yoke through a predetermined amount (typically about 10° of yoke deflection, though it varies somewhat between designs), spoilers will start to deflect on the downward moving wing to increase roll rate and maneuverability. Spoilers can even be used for lift control: on the L-1011, a system called direct lift control (DLC) deflected the four inboard spoilers on each wing 7° up when landing flaps were selected. With DLC operative, a downward pitch movement of the yoke caused the spoiler deflection to proportionally increase drag and cause a greater rate of descent, while an up-pitch input of the yoke caused the spoilers to retract, reducing drag and decreasing the rate of descent. This feature precisely managed the lift of the wing and enabled the aircraft to maintain the same pitch attitude during the entire approach. Although much praised by pilots, it was largely unused in future designs because it is extremely complex to engineer and maintain in a non-fly-by-wire aircraft.

reers, but that is practiced every time they go to the simulator for training). The rudder isn't used to turn the aircraft per se (it can slew the nose of the aircraft around, but it's an uncoordinated and modestly nauseating way to fly the plane), though it does assist in turn coordination, deflecting slightly into the direction of turns when the pilots roll the aircraft. Modern jets have "yaw dampers," which help prevent the tail of the plane from oscillating back and forth, and provide automatic turn coordination during most phases of flight. The results are aircraft where the pilots rarely need to use the rudder pedals except during takeoffs and landings.

Many airliners have no restrictions on flying without their yaw dampers operative. Few passengers would notice, though in some cases those passengers in the rear of the aircraft might notice the tail wagging back and forth to some degree. Many older jetliners (e.g., the Boeing 727) have strict restrictions on operating the plane with inoperative yaw dampers: normally, the plane will have to stay lower and slower to enhance controllability. Special training in the simulator may be mandated for recovering from a yaw damper failure at high speed and high altitude, as the resultant oscillations can be alarming and increasingly difficult to recover from as the event progresses, a trait shared by all varieties of pilot-induced oscillation (PIO).

Airliners have huge control surfaces deflecting into air that is passing quickly. This results in high control forces. Two things make these forces manageable: hydraulically powered controls and trim. Most (but not all) jet airliners have primary control surfaces that are hydraulically powered, while secondary flight controls such as spoilers, speedbrakes, flaps, slats, and trim are either electrically or hydraulically powered (or both). Hydraulics systems are powered by engine driven, electric, or air driven pumps in which hydraulic fluid normally pressurized to 3,000 psi (in some aircraft as high as 5,000 psi) is used to move valves that in turn displace the control surfaces. This is vastly more powerful than a force a human could exert on a surface the size of, say, a 747 elevator; some control surfaces are so large that they can't work without hydraulic pressure. Some smaller (primarily older) airliners (e.g., B-727 and B-737) have a "manual reversion" system where a series of cables and pulleys allow the plane to be flown without hydraulic boost. Having said that, the control forces are extremely high, and both pilots will most likely be physically exhausted by the ordeal. There are numerous variations on the theme: the DC-9 and MD-80 series, for instance, has a hydraulically powered rudder, but the ailerons and elevators are completely manual, using a complex system of cables and control tabs. In most jet airliners, control tabs are for manual reversion only, but in the DC-9 and MD-80 the elevators and ailerons are hinged at the trailing edge of the horizontal stabilizer and wing, respectively, and are not moved using any connection to the cockpit whatsoever; rather, they are positioned aerodynamically by a control tab. This tab is connected to the control yoke (wheel) in the cockpit. When the pilot wants to deflect the ailerons or elevators, the yoke is moved appropriately, and a set of cables deflects the tab opposite the direction the aileron or elevator is to be moved; this force in turn aerodynamically repositions the actual aileron or elevator correctly. It is an elegant piece of engineering that greatly reduces the control forces required and needs no hydraulic power. This is all transparent to the pilot, and serves to introduce another related point of flight control design: trim.

Trim is a key concept that pilots need to master at an early stage of training. Most small aircraft have trim tabs at the back of the flight control surfaces (many only on the elevator), while most jet aircraft achieve pitch trim by means of moving the entire horizontal stabilizer (normally electrically or hydraulically).

The Physics of Trim

A trim tab is a small control surface on the end of a larger control surface (see Figure 3.6). The larger control surface (aileron, or rudder) and the trim tab are little wings creating little lift forces. For example, the aileron creates lift forces that roll the plane about its longitudinal axis, as shown in Figure 3.4B. As the aileron deflection increases, it deflects more air and requires a greater control column force to hold in place. The trim tab, acting as a little wing, creates a force on the end of the aileron that holds the aileron in place without a control column force. The pilot adjusts the aileron (or rudder) and trims the control tab until the pilot input force reduces to zero. Both the ailerons and rudder must be trimmed (because of minute crookedness) to get the plane to fly straight.

The trim tab can be used to adjust the aileron's lift force and thereby alter the position of the aileron. This would normally be done to get the ailerons adjusted to ensure that level flight was achieved with no pilot control inputs. This was available to the Kenya 507 pilots, but the adjustment required was small and the crew was too preoccupied to use it. Strictly speaking, the lift forces are torques relative to a hinge point. The smaller trim tab can cancel the larger aileron with greater leverage. Big plane elevators are a special case. Because high-speed buffet (described in Chapter 4) can affect elevator effectiveness, pitch control for large jets is done by moving the entire horizontal stabilizer. Small planes generally use elevator trim tabs, though some like the Piper PA-28 series adjust the entire stabilizer, just like a jet.

An aircraft in flight will seek a certain speed. At that speed small control movements and light control forces are required to maneuver the aircraft: this is referred to as being "in trim" or "trimmed." A pilot can easily maneuver an 800,000 lb 747 with a feather-light touch of his thumb and forefinger on the control yoke when the aircraft is in trim. The trim tabs work similarly to the control tabs on a DC-9: if you get faster than the trim speed, the plane will start to pitch up to slow down to the speed it is trimmed to. (The location of the trim tabs is shown in Figure 3.6.) If the pilot desires to go faster, he will have to retrim to a new, faster speed. To do that, he will either adjust a trim tab (which will deflect upward, thereby deflecting the elevator downward), the new neutral position of the controls to a faster trim speed, or the entire horizontal stabilizer positions its leading (forward) edge up, with the same net result. If slowing is desired, the opposite occurs. These control inputs may seem onerous but quickly become second nature to pilots. Some more modern Airbus aircraft automatically do this for the pilot at all times ("autotrim"), a feature only possible in highly computerized fly-by-wire (FBW) aircraft. Regardless of how it's achieved, keeping an aircraft in trim is critical to pilots. An aircraft that is far out of trim can be difficult or impossible to control, and procedures for dealing with runaway trim are well ingrained in pilots.

Before takeoff, the horizontal stabilizer trim setting for pitch control is carefully calculated. Each aircraft has a center of gravity (CG) envelope within which all flight must occur (if exceeded, loss of control may ensue), and the stabilizer trim varies substantially depending on multiple variables, including CG, thrust setting, and flap setting. Discussed in Chapter 2 and briefly reviewed here, airliners have a variety of speeds of which pilots are highly cognizant during takeoff. The three most important speeds are V_1, V_R, and V_2 (where V stands for "velocity"). V_1 is the takeoff decision speed: if an engine failure occurs before reaching this speed, the aircraft is guaranteed to be able to stop on the remaining runway if the takeoff is aborted; above this speed, the aircraft has enough speed to successfully fly with one engine inoperative. Think of this

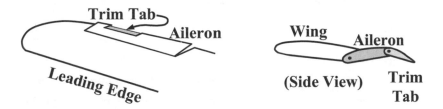

Figure 3.6. Aileron trim tab and enlarged section view.

speed as the speed by which the pilot must begin to stop the aircraft if he wishes to discontinue the takeoff: above this speed, the aircraft is going flying because there may not be enough runway to stop above it. This stop/go decision makes V_1 the most important speed during takeoff. V_R follows (or in some cases can be equal to V_1), which is the speed at which the pilot "rotates" or lifts the nose of the aircraft off the runway. Finally, V_2 is the takeoff safety speed at which the airplane will climb away after an engine failure. Pilots calculate the takeoff trim setting to make the airplane in trim at V_2 so that if an engine failure were to occur during takeoff, they would not be struggling with both an engine failure and an out of trim aircraft, a situation that would tax even skilled and experienced pilots, especially considering the proximity to the ground. There have even been examples of extremely out of trim aircraft being unable to rotate during takeoff.

Further complicating issues in aircraft with engines mounted under the wings of the aircraft (e.g., B-767, B-747, etc.) is that thrust affects the pitch of the aircraft. Because every action has an equal and opposite reaction, when thrust is added in an engine mounted under the wing (and therefore below the CG), it will make the nose of the aircraft rise; conversely, removing thrust makes the nose fall. This is eliminated on Airbus models with autotrim, but on other aircraft, trimming is frequent, especially during approach, when power corrections occur rapidly. It is good that trimming becomes reflexive to pilots; otherwise, pilot-induced oscillation (PIO) incidents would be more common. A typical cycle occurs as follows: a pilot wants to slow the aircraft for landing so he retards the power; this causes the nose to want to drop (on some types the drop is pronounced), which he will counter by trimming nose-up. As the flaps are adjusted and speed is stabilized, he will increase power and the nose will want to rise, requiring him to trim again. On an approach in gusty winds with the airspeed showing nearly constant gains and losses, trimming is almost constant.

One example of the interrelationship of airspeed, trim, and thrust comes from the accident of United Airlines Flight 232, a DC-10, on July 19, 1989. The DC-10 is a three-engine aircraft with two engines mounted under the wings and a third in the tail (above the CG, incidentally). On this flight, the engine in the tail (number 2) failed catastrophically, and shrapnel from its disintegration disabled all three hydraulic systems of the aircraft. The DC-10 has no manual reversion capability, but by judiciously using the thrust from the remaining engines, the crew was able to exert some control over the aircraft and make an emergency landing that saved many lives. When the number 2 engine failed, the aircraft was in trim at cruise at 270 knots. Because DC-10 trim requires hydraulics, that was the speed the aircraft continued to seek for the remainder of the flight, a speed far above any normal landing speed. The aircraft entered a phugoid oscillation, in which the aircraft nose would rise and fall cyclically. In most situations, these oscillations would tend to dampen out over time, but owing to a variety of complex factors in this case did not. The pilots steered the aircraft by using the throttles differentially to turn the aircraft (roll/yaw), and used the throttles together to make the nose rise or fall (pitch). Controlling the aircraft in this scenario seems counterintuitive, as when the nose would fall at the beginning of a phugoid, the airspeed would increase. Because trim was not available, the only means of control that existed to raise the nose was to add power; because the engines were below the CG, the nose would then rise and eventually the aircraft would be back in level, trimmed flight for a brief period before the nose rose too far, speed began to fall, and power had to be reduced to get the nose to come back down. These cycles lasted about a minute each, and was how the aircraft was flown from its cruise altitude of 37,000 feet to the emergency landing in Sioux City, Iowa, in a true feat of airmanship.

Finally, all jet airliners have flaps on the trailing edge (back) of the wing, and most have slats or leading edge flaps (or occasionally both) on the front. Although these devices vary dramatically in their size, capabilities, and complexity, they all serve to alter the flow of the air over the wing, allowing the aircraft to fly much slower during takeoff and landing. Positioning these items is critical, and there are alarms in the cockpit if takeoff or landing is attempted in an incorrect configuration, and procedural and checklist methods in place to ensure that they are properly deployed. Attempting takeoff with flaps set incorrectly is exceptionally hazardous and has resulted in multiple fatal accidents when the takeoff warning system malfunctioned.

Interestingly, the actual amount of flaps used (i.e., degrees of flap deflection) for landing and especially takeoff varies not only between aircraft types, but also on the same type depending on other requirements. Greater flap extension results in more lift, but it also results in more drag. If, for instance, if you needed to depart from a short runway, like at Ronald Reagan Washington

National Airport (DCA, where the longest runway is 7,169 feet in length), you may need a greater flap setting to produce enough lift to get off the ground in such a short distance. The problem is that, once airborne, the aircraft's ability to climb with an inoperative engine (which is always planned for safety reasons) is degraded because of the extra drag produced by the large flap extension used. If a takeoff is limited by the length of the runway, it is referred to as "runway limited"; if it is restricted by the ability to safely climb, it is referred to as "climb limited." The airplane must be under both the runway- and climb-limiting weights (and the maximum structural weight of the aircraft, of course) before it is legal to take off. DCA departures are much more likely to be runway limited.

This section is by no means exhaustive or technically complete, but it provides a basic introduction to flight controls from the perspective of a professional pilot, and should be referenced as necessary during further reading.

The Physics of a Coordinated Turn

If a plane banks its wings and flies in a circle, the water level in a glass inside that airplane tilts, as shown in Figure 3.7B. The so-called centrifugal force shoves the water sideways to the outside edge. In a "coordinated turn," the water level in the glass, passengers, and plane all tilt together, as shown in Figure 3.7C. Unless the passengers look out the window, they are unaware that the plane is turning. A coordinated turn without any apparent sideways shove is favored for passenger comfort. Our goal here is to define the flight parameters (i.e., bank angle, speed and turning radius) required for a coordinated turn. This effect contributed to the Cameroon accident pilots apparently being unaware their plane was banking and turning.

A centripetal, or center-seeking, force is required to create circular motion. Consider rotating a weight on the end of a cable. The cable exerts a force on the weight, forcing it to deviate from a straight path. The cable force constantly changes direction but always points toward the center of rotation. (The cable force substitutes for the horizontal component of lift when the plane banks its wings.) The inertia of the plane's mass balances the cable force per $f = ma$ (with acceleration = velocity^2/radius). Instead of swinging a weight on the end of a cable, we will consider the forces on a

 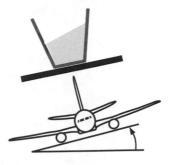

(A) Water in Glass: Level Flight   **(B) Uncoordinated Turn**

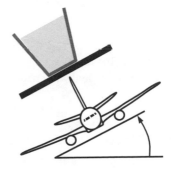

Figure 3.7. Effect of a coordinated turn on water level in a glass.

(C) Coordinated Turn

150,000 lb Boeing 737 flying in a circle. A big jet typically completes a 360° turn in about four minutes. An example holding pattern speed might be 250 knots. Completing a 360° turn in four minutes at 250 knots works out to a circle with a radius of 3.05 statute miles. Our imaginary 737 does not have wings but instead slides in a frictionless horizontal plane tethered to a 3.05-mile-long weightless cable. Our wingless 737 is self-propelled at a constant 250 knots. The cable is held taut by hanging a weight suspended over a pulley, as shown in Figure 3.8. Two additional scientific principles are required: the acceleration of circular motion and the inertia effect created by that acceleration.

Normally we think of acceleration as the rate of change of velocity; a classic example is the proverbial falling apple discussed in Chapter 1. Ve-

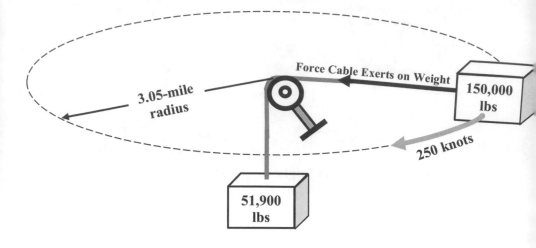

Figure 3.8. The forces acting on a Boeing 737 flying in a circle. This imaginary 737 is sliding on a frictionless horizontal plane and a tethered to a cable held taut with a weight.

locity, however, is a vector that has both direction and magnitude. Even though the plane is flying at a constant speed, the velocity vector is constantly changing directions. This changing velocity vector direction defines a rate of change of velocity (i.e., acceleration). This is best understood by first understanding the concepts of vector addition and subtraction.

A plane flies 400 miles due north, turns, and then travels due west for 100 miles. Displacement relative to an origin is a vector. The initial displacement vector has magnitude (400 miles) and direction (due north). Adding two vectors is more complicated than adding two numbers because the direction and magnitude must both be considered. The result can be calculated precisely with the Pythagorean theorem to be 412.3 miles, with orientation as shown in Figure 3.9. An approximate result can also be obtained with a carefully scaled drawing. Accuracy improves with a larger drawing.

Velocity, the rate of change of displacement, is also a vector. If the plane is flying 400 knots (magnitude) due north (direction) while simultaneously experiencing a 100-knot jet stream due west, the net effect is that the plane is flying 412.3 knots with a heading 104° counterclockwise from the x axis. If we relabel "miles" in Figure 3.9 as "knots," we can use the same graph to determine the resultant velocity vector. The displacement vectors 400 miles north and 100 miles west vectorially add the same as velocity vectors 400 knots north and 100 knots west.

Consider a plane flying in a circle, as shown in Figure 3.10. The velocity at some instance of time is V_{1C} (subscript C is added to avoid confusion with previously defined takeoff speeds). After an incremental passage of time Δt, the velocity is now V_{2C}. V_{1C} and V_{2C} have the same magnitude (the plane is flying at a constant speed); however, the changing direction creates a change in the velocity vector. If we locate V_{1C} and V_{2C} at the same point, we can write the vector equation $V_{1C} + \Delta V = V_{2C}$ (see Figure 3.10) or, using the concept of vector subtraction, $\Delta V = V_{2C} - V_{1C}$. Even though V_{1C} and V_{2C} have the same speed, ΔV represents the vector subtraction of V_{1C} from V_{2C}. Acceleration is the rate of change of the velocity vector and is equal to $\Delta V/\Delta t$. Because Δt is a number (i.e., not a vector), the acceleration vector $\Delta V/\Delta t$ has the same direction as ΔV. As seen in Figure 3.10, ΔV doesn't exactly point toward the center of rotation but would if we considered a small increment of time. With a carefully scaled drawing, we could estimate the centripetal or center-seeking acceleration of the plane flying in a circle, and we would

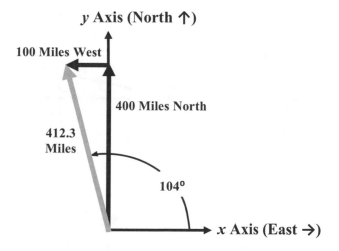

Figure 3.9. Resulting displacement of a plane that flies 400 miles north followed by flying 100 miles west. The result is the vector sum of two displacement vectors.

discover that this acceleration equals the plane's speed squared divided by the radius of the circle. Velocity²/radius, or V^2/r, is the standard equation for acceleration during circular motion at constant speed. The V^2/r acceleration for the cable-tethered vehicle traveling 250 knots (422 ft/sec) at a radius of 3.05 statute miles (about 16,100 feet) works out as 11.06 ft/sec².

Acceleration creates an inertial response. Inertia is a property of matter that resists acceleration according to Newton's law $f = ma$. Inertial loads are more easily visualized by standing on a scale in an elevator. If the elevator is accelerating up at 16 ft/sec² (one-half of gravity's normal acceleration), a scale under a person weighing 100 lbs will read 150 lbs: 100 lbs of normal weight and 50 lbs of inertial loading. The inertial load acts like a force, and it can be read on the scale like a force. It's not really a force but rather the effect of the person's mass resisting acceleration per $f = ma$. The person accelerating at one-half of normal gravitational acceleration will feel as though they weigh 150 lbs. The inertial load always acts in the opposite direction of the acceleration. The elevator accelerates up; the inertial effect is down and increases the reading on the scale.

The force under the scale required to accelerate the 100 lb person up at 16 ft/sec² can be calculated with Newton's law; the sum of the forces in the

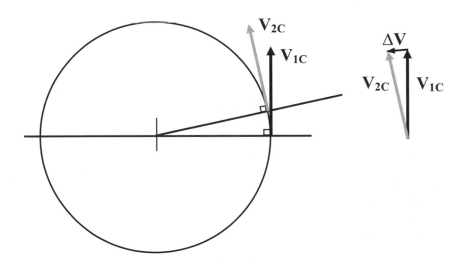

Figure 3.10. The velocity vectors of a plane flying in a circle at constant speed.

vertical direction equals mass × acceleration. (Don't forget to convert the person's weight to mass by dividing by the acceleration of gravity, as explained in Chapter 1.)

$$\text{Force} - 100 \text{ lbs} = \text{mass} \times \text{acceleration} = \frac{100 \text{ lbs sec}^2}{32 \text{ ft}} \times \frac{16 \text{ ft}}{\text{sec}^2}$$

The required force is found to be 150 lbs. This force is balanced by the person's 100 lbs of weight and 50 lbs of inertial loading. Accelerating at one-half of the normal gravitational acceleration creates a downward inertial load equal to one-half of the person's weight. Accelerating at any percentage of the normal gravitational acceleration (32 ft/sec²) creates a downward inertial load equal to the same percentage of the body's weight.

The inertial load for circular motion is sometimes called the centrifugal force. This is a misnomer because strictly speaking it is not a force but instead an inertial effect—the mass × acceleration term in Newton's equation. For the 150,000 lb plane flying in a circular holding pattern (radius 16,100 feet) at 250 knots (422 ft/sec), the acceleration pointing toward the center is 11.06 ft/sec². The inertial load is opposite the direction of the acceleration (pointing away from the center of the circle). The inertial load equals 11.06 ÷ 32 = 0.346, or 34.6% of the plane's weight equal to 0.346 × 150,000 lbs = 51,900 lbs, which is the weight used in Figure 3.8 to balance the wingless plane tethered to a cable.

And of course the actual plane has wings; the 51,900 lbs are supplied by the horizontal component of lift, as shown in Figure 3.11. In level flight, the vertical component of lift must always equal the plane's weight. The correct bank angle that makes everything balance (for a given set of conditions, e.g., a 150,000 lb plane moving at 250 knots with a turning radius of 3.05 miles) is the angle that creates a horizontal component of lift equal to 51,900 lbs, or a horizontal component of lift that is 34.6% of the plane's weight. This angle can be determined from a carefully scaled drawing, such as shown in Figure 3.11, or directly calculated with trigonometry to be 19°.

If a 100 lb passenger is sitting in a plane that is moving at 250 knots with a 3.05-mile turning radius, there is an inertial load of 34.6 lbs, a vector that adds to the person's weight vector, as shown in Figure 3.12. The angle of the

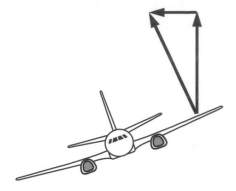

Figure 3.11. Forces acting on a turning plane. The vertical component of lift must equal the plane's weight. The horizontal component of lift creates circular motion.

resultant vector sum can be determined from a carefully scaled drawing or precisely calculated to be 19°. If a plane traveling 422 ft/sec in a circle of 16,100 feet and banks its wings 19°, the result is a coordinated turn. The resultant vector sum of the inertial load and weight load pass through the passenger's spine, there is no sideways shove, and a glass of water appears to the passenger to remain level. The vector sum of the passenger's weight and inertial loading is the square root of $(0.346^2 + 1^2) = 1.058$. The 100 lb passenger, if sitting on a scale, will appear to weigh 105.8 lbs when turning as described.

Because of the added inertial load, the passenger appears to weigh 5.8% more, and so does the plane. To maintain level flight, the wings must supply 5.8% more lift by increasing the wing's angle of attack. (Excess angle of attack will eventually create an aerodynamic stall, a topic discussed in Chapter 4.) There are mandated requirements for wing strength. The Federal Aviation Administration requires the wings to withstand a load factor of 2.5 with a safety factor of 1.5 before failure. A load factor of 2.5 corresponds to a bank angle of 66°. Pilots are taught to never exceed a bank angle of 30°.

During certification testing, the wings are loaded with hydraulic actuators until they support 1.5 × 2.5 = 3.75 times the maximum weight of the plane. The Boeing 777, for example, completed certification testing in 1995 with wingtip deflections of 24 feet before the wings snapped! (A few feet of wingtip bounce during turbulent flight is harmless.) The Airbus A380

failed just short of the 150% safety margin during its wing test in 2006. Instead of retesting (and destroying a second plane worth hundreds of millions of dollars), European regulators allowed Airbus to submit computer
simulation demonstrating that 66 lbs of added reinforcement will meet all
requirements.

A load factor of 3.75 could occur during pulling out of an emergency dive
or flying a coordinated turn with a bank angle of 75°. If a plane reaches a
75° bank angle, clearly an upset condition, the wings do not necessarily fall
off because the plane is twisted around in space and not flying a continuous
coordinated turn in a level geometric plane. The ultimate proof of adequate
design standards: the wings have never fallen off a large commercial passenger jet in the modern era. As another example of wing load safety margins, in 1965, a Boeing 707 landed safely in spite of losing 25 feet of wing
after an engine fire. Obviously, this unbalanced plane would roll; the pilot
corrected with control surfaces.

Does the pilot have to struggle with the three variables (bank angle, speed,
and radius of curvature) to coordinate the turn? No! Controllers assign an
altitude to fly a holding pattern; the pilots will fly the most fuel-efficient

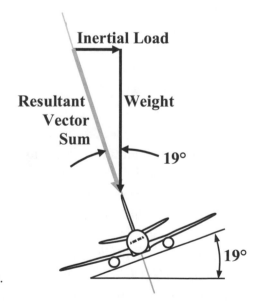

Figure 3.12. Vector sum of
inertial load and weight load for a
coordinated turn at 250 knots and
turning radius of 3.05 statute miles.

Spatial Disorientation

Spatial disorientation, a complex subject involving many senses, consists of the body's ability to sense motion and orientation. Early on, it was determined that experienced pilots could not fly straight and level without seeing the horizon, which necessitated the development of the artificial horizon in 1929. Part of the problem is the difficulty of sensing turns. If the plane banks 30°, the net effect is that the pilot gradually weighs 15% more, which is difficult to sense and a totally different effect than when driving a car. When turning in a car, inertial loading shoves the driver sideways to the outside of the curve. Inability to sense turns is believed to be a factor in loss of control (LOC) accidents that typically occur when the horizon is not visible. Additionally, as mentioned in the Kenya 507 accident discussion, rates of movement below a certain threshold cannot be detected by the human vestibular system, allowing incorrect perceptions of spatial location to gradually lead to unintended aircraft attitudes when there is no visible horizon. Boeing and most airlines define an upset condition as a pitch up of 25°, a pitch down of 10°, or a bank angle of 45°. Beyond these limits, the aircraft needs to be urgently recovered to prevent an LOC accident. Upset conditions (or "unusual attitudes," as they are sometimes called) are frequently a by-product of spatial disorientation.

speed for the plane type and altitude (but can't exceed the maximum holding speed for that altitude). Planes fly faster at higher altitudes, and the protected airspace at higher altitudes is correspondingly larger. The pilot banks the plane, which automatically establishes a radius of curvature that balances the so-called centrifugal force with the horizontal component of lift. Federal Aviation Administration rules govern holding patterns; planes hold at approximately 30° of bank at the ends of the holding pattern, adjusted for winds. The result is that planes of all sizes are turning with the same radius when flying at the same altitude and airspeed. Pilots and controllers don't have to concern themselves with turn radius: so long as the plane is at or below the maximum holding speed for their given altitude, the holding pattern will never be larger than the protected airspace designated for the hold.

To demonstrate the effect of changing bank angle, we go back to Figure 3.8 and double the hanging weight. If the suspended weight is suddenly doubled to $2 \times 51{,}900 = 103{,}800$ lbs and the speed remains constant, the additional weight pulls on the cable and reduces the radius. Its $f = ma$, with ma equal to mV^2/r. The radius reduces until the 103,800 lb weight is in equilibrium with mV^2/r. This occurs with a new radius of 8,042 feet (1.523 miles).

Straight and level flight can be considered turning with an infinitely long turning radius. As the plane begins to bank just a little, the turning radius becomes a large finite number; the circular acceleration and inertial loading is small. The turning radius automatically decreases as the plane increases the bank angle, analogous to increasing the hanging weight on the vehicle tethered by a cable.

Hanging a heavier weight from the cable is analogous to increasing the bank angle and increasing the horizontal component of lift. For the airplane, the horizontal lift component doubles when the bank angle increases from our original 19° to 34.68°. Assuming level flight (not climbing or descending), flying a coordinated turn with a 1.523-mile radius at a speed

Flying with One Wing

The vertical component of lift vanishes as the roll angle approaches 90° and the load factor becomes infinite. (The load factor for level flight with roll angles of 80° and 89° is 5.75 and 57, respectively.) Acrobatic and fighter planes routinely fly sideways (i.e., 90° roll) with no wing lift supporting the plane. Lift is created along the fuselage by pointing the nose slightly up. Also, there is a vertical component of engine thrust. In fact, damaged fighter planes missing a wing have been known to fly a 90° roll and level the remaining wing before landing safely.

of 250 knots and a bank angle of 34.68° requires a load factor or total lift on the wings 1.216 times the weight of the plane.

How could the Cameroon accident pilots not notice that the plane was banking and turning until the bank alarm sounded at 35°? Because the plane was in a coordinated turn, the water level in the glass appeared to remain level. The pilots gradually weighed 20% more, but this inertial load is applied directly through their spines with no side thrust. The distracted pilots did not even notice their plane was turning with an increasing bank.

Vertical Stabilizer

The plane's forward thrust creates a forward $f = ma$ acceleration. This acceleration increases the plane's speed until the drag force equilibrates with engine thrust. When the forward thrust equals the drag force, the acceleration stops and the plane continues forward at a constant speed. The horizontal lift force creates the same effect; the plane accelerates sideways until the sideways drag force cancels the horizontal lift force, and the plane now has a constant horizontal speed. If no other forces act on the plane (an incorrect assumption), the plane does not fly in a curve but instead flies off at an angle, as shown in Figure 3.13.

The plane with banked wings does not fly in a straight line at an angle; it flies in a circle. There must be an additional force on the plane that continuously changes the direction of the horizontal lift force. This additional force yaws the plane and keeps the horizontal lift force continuously pointing toward the center of rotation.

The cross-section of the vertical stabilizer is shaped like a symmetrical wing. Because of symmetry, there is no net lift on the vertical stabilizer when the plane flies straight and level. When the plane banks its wings and tries to fly sideways, the relative air flow over the vertical stabilizer creates a lift force that continuously yaws the plane, as shown in Figure 3.14. The net effect is that the plane flies on a curved path. How does the tail know how to yaw the plane the correct amount? The vertical stabilizer keeps the plane aligned with circular motion; that is, the longitudinal axis of the plane (the axis along the fuselage centerline) remains perpendicular to the radius of curvature. If the nose points toward (or away) from the center of rotation, asymmetric drag on one side is self-correcting and keeps the nose

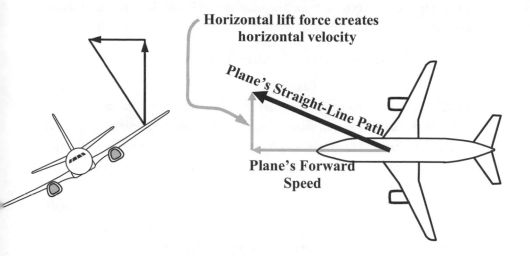

Horizontal lift force creates horizontal velocity

Plane's Straight-Line Path

Plane's Forward Speed

Figure 3.13. If the only forces acting on the plane are the horizontal component of lift, the plane's forward velocity vector adds to the horizontal velocity vector. The plane does not fly in a curve but instead flies in a straight line on an angle. An additional force is required to create curved flight.

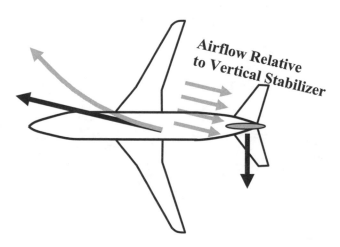

Airflow Relative to Vertical Stabilizer

Figure 3.14. The straight line shows the plane's path if there are no forces on the vertical stabilizer. The vertical stabilizer force constantly yaws the nose of the plane to the pilot's right, creating a curved path.

pointed in the correct direction. The nose points in the direction that minimizes drag and maintains symmetric drag forces.

Big Planes Have Big Vertical Stabilizers

The rudder in the tail must balance the plane during any engine out scenario. (With one engine out, the unbalanced plane yaws or rotates about the vertical axis. Rudder action rebalances the plane and prevents rotation.) As engines became pod mounted and hung farther from the fuselage, a bigger rudder and correspondingly bigger vertical stabilizer were needed to balance the engine-out scenario.

Historically, piston engines were always mounted in the wing. The de Havilland Comet, the first commercial jet, continued this trend. The vertical stabilizer for a contemporary Boeing 737 and de Havilland Comet is shown approximately to scale in Figure 3.15 (the Comet's wingspan was 114 feet 8 inches vs. 94 feet 9 inches for the 737). Also shown is a small Cessna with a centered engine. The single-engine Cessna does not need a large vertical stabilizer to balance the one-engine-out scenario. A slightly larger two-engine Cessna requires a larger vertical stabilizer.

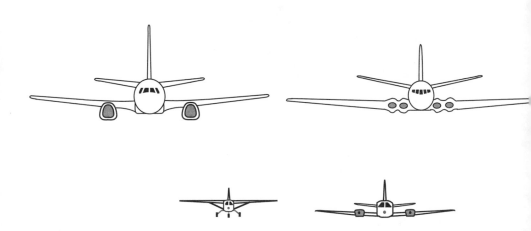

Figure 3.15. Compare the relative height of the vertical stabilizer for a contemporary Boeing 737, a wing-mounted de Havilland Comet, a single-engine Cessna, and a two-engine Cessna.

Pod-Mounted Engines

Boeing invented pod-mounted engines at the request of the US Air Force on the first jet-powered bomber, the B-47. The Air Force required engines to be located as far as possible away from the pilots and bombs (engines can spin themselves apart; see Chapter 5 in *Beyond the Black Box*). Hopefully providing isolation of engine fires, the pod-mounted engine was a significant safety improvement. Hanging the engines farther out strengthens the wing; the engine's downward weight somewhat cancels the upward lift force.

Large commercial planes, with their large vertical stabilizers, turn without the pilot using the rudder (in actuality, the aircraft's yaw damper may use small amounts of rudder to automatically coordinate the turn, but this is transparent to the pilot and does not move the rudder pedals). Large commercial jets weathervane; that is, the plane aligns itself with the circular motion. Small planes, with small vertical stabilizers, must use the rudder coordinated with the bank angle to correctly turn the plane. With too much or too little rudder, the plane is flying a bit sideways relative to circular motion (i.e., nose pointed toward the center of rotation or away). The sideways motion increases drag and wastes fuel.

Captain Hedges Explains: What Is Coordinated during a Coordinated Turn?

A coordinated turn occurs when the correct amount of rudder is being applied while turning; that is, there is no sideslip on the turn and bank indicator (or on some aircraft the Beta target, where Beta is the aerodynamic label for yaw, which the rudder provides). In small aircraft, this is a significant challenge because the control surfaces are all mechanically controlled (and propellers have several destabilizing tendencies). Modern jets are stable, but it is possible for a pilot to misuse the rudder, an extreme instance of which occurred on November

12, 2001, when the First Officer of American Airlines Flight 587 (an A300) rapidly applied rudder in alternating directions, eventually ripping the rudder and vertical stabilizer off the aircraft, leading to a crash. In the case of Air Transat 961 (see page 130), the A310 had an undetected defect in the composite structure of the rudder, which caused the rudder to detach in flight; in this case, the crew experienced Dutch roll, which abated as the aircraft descended into denser air (see the section on "Dutch Roll" on page 126). Unlike the American A300, the Air Transat aircraft retained its vertical stabilizer, which allowed the pilots sufficient controllability to make a safe landing in Cuba. These two accidents led to regulatory changes; in the case of American 587, pilot training was significantly improved with regard to rudder use and limitations, while Air Transat 961 led to changes in inspection protocols for composite parts on aircraft.

Big planes have vertical stabilizers bigger than needed for a coordinated turn because they are sized in case an engine goes out. Because it is bigger than normally needed, airliners tend to need little rudder to remain coordinated because that is the orientation that minimizes drag. Staying coordinated is important for several reasons. First, it makes for a much more comfortable flight for the passengers, and second, it ensures that the airplane is going through the air in the most efficient way possible; if either too much or too little rudder is used, a larger portion of the side of the aircraft is being exposed to the airstream than necessary, which produces more drag and poorer performance. Proper rudder management is critical in keeping the aircraft coordinated after an engine failure, particularly at slow speeds, such as right after takeoff. Failure to properly use the rudder can have dire consequences on the aircraft's ability to climb in that situation, which is one reason airline pilots extensively practice this maneuver every time they go to the simulator for training. A particularly challenging scenario that pilots of three- and four-engine aircraft train for is an approach and landing with two inoperative engines. For four-engine aircraft, the two engines must be on the same wing, while for the three-engine aircraft, the center and one other engine must be inoperative to ensure that the pilots proficiently demonstrate the maximum yaw corrections possible. These are challenging maneuvers from which a go-around may not be possible: in most aircraft types, the pilots are committed to land after extending the landing gear.

Captain Hedges Explains: Holding Patterns

A common application for a coordinated turn is flying the turns at the end of a holding pattern. Holding patterns are racetrack-shaped patterns in the sky designed to allow aircraft to delay their flights in predefined airspace. Holding patterns can be assigned for a variety of reasons, but the two most common reasons are severe weather at the destination airport or an exceeded maximum arrival rate, where air traffic control (ATC) must delay certain flights to allow for adequate spacing between aircraft. Holding patterns are defined relative to a certain geographic point called the holding fix. Some holding patterns are "charted" (i.e., defined on a map), though ATC can make up their own. In addition to the assigned holding fix, the aircraft assigned to hold will be told what bearing or radial from the holding fix they are to hold on and will be given an expect further clearance (EFC) time, which serves two purposes. First, it helps the pilots know how long they can be expected to hold and determine whether they have adequate fuel to hold for that long (many diversions occur this way). Second, if the aircraft were to have a radio failure during the hold, the EFC time would be when they were expected to leave the holding pattern and continue their arrival. By default, holding patterns are to the right, but controllers can specify left turns if need be. Likewise, the length of the inbound leg of the holding pattern is 1 minute at and below 14,000 feet and 1.5 minutes above 14,000 feet. In the United States, the length of the outbound leg must be adjusted for winds to ensure that the inbound leg is the correct length. Additionally, ATC or charted holding patterns may have leg length specified by distance rather than time; if a hold is planned to be quite long, pilots will frequently request longer leg length for two reasons: (1) it's more efficient, as the aircraft requires more power to sustain its speed in the turns at the ends of the pattern than it does when straight and level on the outbound or inbound legs, and (2) passengers find straight and level flight to be more comfortable than frequent turns. For reasons of efficiency, it is important that the aircraft be properly trimmed and coordinated throughout the holding pattern.

There are different maximum speeds for holding patterns, depending on the plane's altitude. From the ground up to 6,000 feet, the maximum speed is 200 knots; above 6,000 feet up to 14,000 feet, the maximum speed is 230 knots; and above 14,000 feet, the maximum speed is 265 knots. These speeds and leg lengths are specific to the United States, but many other countries have their own holding nuances; it's up to the pilot to study any differences in procedure before flying internationally.

In practicality, every aircraft has a best holding speed; in airliners, it depends primarily on weight and altitude and is normally close to L/D_{MAX}. Three minutes before getting to the holding fix, the pilot should slow to best holding speed. On some airliners, the best holding speed may be above the maximum holding speeds listed above, in which case they will either have to ask ATC if they can hold faster or extend flaps to allow them to comply with the maximum speed (this is not preferred, as it burns much more fuel and may result in a more rapid diversion to an alternate airport). Typically, modern jets can enter and fly the holding pattern using the autopilot, but pilots are required to remain proficient in holding patterns and entries.

A typical holding pattern clearance would sound something like, "Cleared to the Charlotte VOR, hold south on the 174° radial, left-hand turns, 10-mile legs, expect further clearance at 14:42 Zulu, maintain 10,000 feet." This holding pattern is shown in Figure 3.16. Let's break this down a bit. The first piece of information is where the plane will be holding, in this case, the Charlotte VOR (a type of radio navigational aid). The controller then said that the crew should hold south on the 174° radial, which is a line sticking out of the Charlotte VOR (the 180° radial would be directly south of the station). The pilots will mentally add 180° to the radial to find that the inbound course to the Charlotte VOR is 354° (most newer aircraft can deduce this if entered correctly into the flight management computer by the pilot). After the flight gets to the VOR, ATC wants the flight to make the holding pattern turns to the left. In this case, they told the flight that they could expect to come out of the hold at 14:42 (2:42 p.m.) Zulu time, which is defined as the time in Greenwich, England, the accepted time base for flying (and which is five hours ahead of the time in Charlotte, North Carolina, during the summer, meaning more math for the pilots to do). Finally, the crew was instructed to maintain 10,000 feet (which could have also been given at the beginning of the clearance), which means that the crew will have to slow to a maximum of 230 knots prior to crossing the Charlotte VOR because they are between 6,001 and 14,000 feet. The crew must be mindful of where they are, where they are going, and how they are going to get into the holding pattern (there are three different entries depending on which direction the holding fix is approached from, but they are beyond the scope of this book). They must adjust their speed prior to the holding fix, comply with the holding pattern instructions, and determine how long they can hold before they need to divert to an alternate airport. As a practical matter, they will then notify their company dispatcher of the hold, the EFC, and fuel on board, and will finally tell the Flight Attendants and make an announcement to the passengers. Last-minute holding clearances can be extremely challenging, which is why pilots are frequently tested on them.

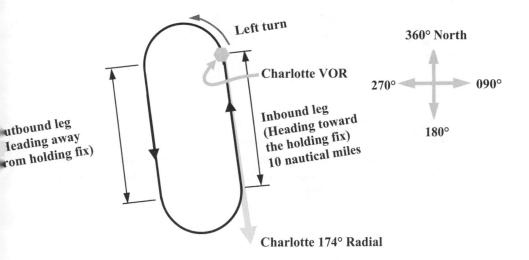

Figure 3.16. Holding pattern for Charlotte. The course on the inbound leg of the holding pattern is 354°.

Dutch Roll

The basic motions of roll, pitch, and yaw can adversely interact in a variety of ways. One such interaction is famous enough to have its own name: Dutch roll. This motion is like that of a Dutch speed skater who rolls and yaws when swinging his arms back and forth.

Shortly after the end of World War II, engineers raced to Germany searching for aviation secrets. One such discovery was swept wings. Swept wings greatly reduce drag at high speeds by delaying shockwave formation.

Flying above the speed of sound creates shockwaves, as the plane is moving faster than the air can get out of the way. Forming a shockwave requires energy and therefore increases drag. Because airflow over the top of the wings is faster than the airplane's speed, the shockwaves are formed even when flying at subsonic speeds. For that reason, transonic planes—those that fly at 80% to 100% of the speed of sound—have swept wings. At sea level, the speed of sound is 767 mph. The speed of sound varies with temperature and decreases to about 660 mph at 40,000 feet.

In 1945, Boeing was designing a large jet-powered bomber—the B-47. After the German discoveries mentioned above, the initial straight wing (i.e., 0° swept back) was altered to a 35° swept-back design.

During test flights (1948), the B-47 would not fly straight on its own! If the plane yawed for any reason (turbulent gust or pilot oversteer), Dutch roll was triggered. Worse still, the B-47's yawing and rolling oscillations increased with each cycle. After yawing, the forward wing sees more direct airflow and has greater lift. The lift unbalance causes the plane to roll, as shown in Figure 3.17.

The forward wing also experiences more drag. The increased drag yaws the plane back, and if the plane overshoots and yaws in the opposite direction, the process repeats. The plane experiences a coupled yawing and rolling motion.

Boeing solved the B-47's problem with a "yaw damper," a device that sensed the onset of Dutch roll and automatically applied a counter-rudder movement to reduce the yawing motion. Boeing's first commercial jet, the 707, also had 35° swept wings, and it also had a Dutch roll problem.

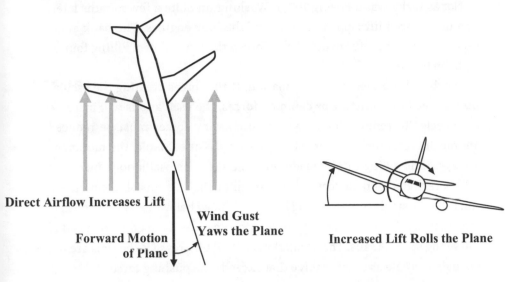

Direct Airflow Increases Lift

Forward Motion of Plane

Wind Gust Yaws the Plane

Increased Lift Rolls the Plane

Figure 3.17. Plane yaws from a wind gust. A forward wing experiences more direct flow and increased lift that rolls the plane.

Certification of a new plane design involves rigorous test flights. Certification also defines pilot training requirements. The FAA, for the first time, required Dutch roll recovery training for Boeing 707 pilots. This involved shutting off the yaw damper (to mimic equipment failure), purposefully inducing Dutch roll and performing Dutch roll recovery. Pilots can learn, with practice, to manually reduce the oscillations with proper timing of control surface inputs; unfortunately, incorrect timing can make the oscillations worse.

Tex Johnston, the Boeing test pilot famous for doing a barrel roll[4] (in 1955 in a 707 prototype), states that 707 training stressed a maximum yaw of 15° during Dutch roll recovery. Excess yaw during a 707 training flight near Paris in 1959 caused unbalanced lift so violent that centrifugal forces ripped an engine from its mounts. (Engines are mounted with mechanical fuse pins designed to cleanly break away and protect the wing.) The damaged 707 managed to safely land.

Not as lucky was a Boeing 707 in Washington State a few months later. The unbalanced lift ripped out three of the four engines. The bank angle reportedly reached 40° to 60°. A fire forced the plane down, killing four of eight on board.

Dutch roll is a mechanical oscillation. If the forces creating the oscillations exceed the resisting or damping forces, the oscillations will increase each cycle. To reduce Dutch roll, the oscillating forces must be reduced and/or the resisting forces must be increased. For example, the motion of a playground swing can be made unstable (i.e., the oscillations increase) with a repeated and correctly timed small push. This unstable pendulum can be made stable if the pushing force is reduced below the system's resisting frictional force. The swing's oscillations can be damped or reduced if the swing sprouts sails. The oscillations will not increase if the sails are big enough to create a resisting force that exceeds the pushing force.

After the accident, Tex Johnston spent two days reviewing previous near-catastrophic Dutch roll recovery incidents during training flights. (No incidents or accidents occurred with passengers.) Johnston decided that the 707 required design changes.

Like the swing sprouting sails, Tex Johnston recommended increasing the size of the 707 vertical stabilizer to increase yaw resistance. The height of the tail grew by 35 inches, and a 39-inch ventral fin was added to the bottom of the fuselage (directly underneath the vertical stabilizer).

The problems decreased but did not completely go away. Captain Hedges reports that the Boeing 727, Boeing's next commercial plane with a wing sweep of 32°, was challenging to recover from Dutch roll with the yaw dampers shut off, and demonstrating Dutch roll recovery during simulator training was required for pilot certification to fly the type.

Boeing's next commercial plane, the 737, is dynamically stable with respect to Dutch roll. That is, the oscillations automatically reduce without any pilot inputs. Dutch roll oscillations decrease with decreasing sweep angle. The 737's reduced sweep angle of 25° is most likely related to its Dutch roll dynamic stability.

By reducing the oscillating forces and increasing the resisting forces, Dutch roll has been designed out of modern commercial planes. The yaw damper still exists, largely to prevent passenger motion sickness, and is no longer considered critical for flight safety.

Dutch roll recently reared its ugly head and caused a US Air Force Boeing KC-135 Stratotanker to crash in Kyrgyzstan in 2013 just 11 minutes after takeoff. The KC-135 (produced from 1955 to 1965) is a military version of the 707 and is used as a flying tanker to refuel planes in flight. The Stratotanker has the 707's wing sweep angle and a yaw damper critical to safe flight. During the accident flight, the yaw damper failed. Pilot rudder inputs made the oscillations worse. Full rudder deflection with excessive yaw causes the rudder to scoop out or resist more airflow than permitted by structural limits. The tail section broke off and the Stratotanker went into a dive, killing all three crew.

The military investigators concluded that the flight crew lacked adequate training for Dutch roll recovery. (The flight control augmentation system, a newer feature, made the latest version of the KC-135 more stable. Crews no longer practiced Dutch roll recovery.) Surprisingly, the problem could not be duplicated in the simulator. So the simulators and training procedures were improved.

In a related problem, airplanes are not designed to fly sideways. Full rudder deflection induces excessive yaw and can eventually do structural damage to the tail as the rudder scoops out more air than designed. This creates excess structural loads, so most airliners have rudder limiters installed to prevent pilot rudder inputs large enough to damage the aircraft.

The structural failure that occurred with the KC-135 in 2013 was similar to a 2001 Airbus crash. In both cases, large rudder deflections with excess yaw caused structural failure of the tail section. Alternating rudder inputs present the largest hazard and are not demonstrated in aircraft certification programs; this was the specific cause of the 2001 American Airlines A300 accident, and since then pilot training syllabi have been revised extensively to train pilots not to make large alternating rudder movements.

Dutch roll can be controlled with the rudder, ailerons, or a combination of the two. Because of the possibility of structural damage associated with excess yaw and large rudder deflections, Dutch roll control with ailerons is the recommend method.

Air Transat Flight 961

In yet another surprising occurrence of a problem thought to be obsolete, a modern Airbus A310 experienced Dutch roll. The plane departed Cuba bound for Quebec on March 6, 2005, with 271 passengers and crew on board.

At about 90 miles south of Miami and an altitude of 35,000 feet, the crew heard a loud bang followed by a few seconds of vibrations. Flight Attendants in the back of the plane were thrown to the floor as the airplane started to Dutch roll.

The pilots disconnected the autopilot and began to manually fly the plane. The airplane was difficult to control. To reduce pilot workload, the second autopilot system was engaged. Dutch roll intensified, and the pilots again reverted to manual flight. The crew reported problems with both autopilots.

Knowing that Dutch roll would likely reduce and stop at a lower altitude (the denser air better resists the yawing motion), the pilots descended the plane. Trying to diagnose the problem, the crew cycled through the electronic centralized aircraft monitoring system, without any results. Dutch roll stopped when the plane descended through 28,000 feet. The plane turned around and landed uneventfully in Cuba.

What induced Dutch roll, a problem designed out of modern aircraft? If the Boeing 707 can significantly reduce Dutch roll by increasing the size of the vertical stabilizer, perhaps Dutch roll can suddenly appear on an Airbus A310 if the vertical stabilizer is suddenly smaller. Because of a unique manufacturing defect, the carbon fiber–reinforced plastic rudder suddenly broke off. The absence of the rudder had little effect on control of the plane, but the decrease in the vertical area of the rudder allowed Dutch roll to occur.

4

Vanished!

Quickly becoming the greatest mystery in aviation history, Malaysia Airlines Flight 370 took off on March 8, 2014, at 12:40 a.m. local time from Kuala Lumpur, Malaysia. The Boeing 777, with 239 passengers and crew, was bound for Beijing. Following standard procedures, the plane had enough fuel to fly about 2 hours beyond the scheduled 5.5-hour flight.

Last voice contact occurred at 1:19 a.m. Numerous communication systems were either shut off or somehow failed, beginning at 1:07 a.m. Malaysian military radar recorded the plane turning right followed by a sustained left turn at 1:21 a.m. The plane made additional minor changes in direction and altitude followed by another major turn (see Figure 4.1) before disappearing from military radar at 2:22 a.m.

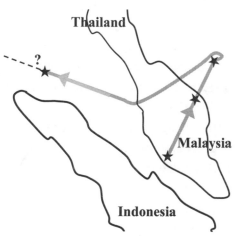

Figure 4.1. The stars indicate (in direction of flight) takeoff, last message sent by the Aircraft Communications Addressing and Reporting System (ACARS), last trans-ponder message, and the last military radar contact of Malaysia Airlines Flight 370.

Planes with unconscious crews, which do not change direction or shut off communication equipment, have flown for hours.[1] A fire can progressively knock out electronics, but why not the flight computers? If the plane is on fire, the pilots, seeking clearance to the nearest airport, radio controllers, or if over water and ditching becomes necessary, the flight crew radios current position and status details in preparation for rescue. (If a plane enters the water under pilot control, survivors are expected). Though hypoxia and/or fire can explain some of the features of the missing plane, none of these scenarios seemingly account for the loss of Malaysia 370.

Controllers (Sort of) Monitor Location

Shortly before World War II, radar was developed in several countries. Primary radar transmits radio pulses and detects a significantly weaker reflected signal. A powerful transmitter is required. The process is also susceptible to electronic noise, dispersion, refraction, and spurious reflections.

Air traffic controllers use primary radar but also send an interrogation signal. Each plane's transponder sends out a radio transmission that responds with more detailed information that generates a much more comprehensive picture for the air traffic controller on a computerized secondary radar display. Two approaching planes also use the transponder signal to trigger collision avoidance systems. In the United States, Federal Aviation Administration (FAA) rules require all planes flying above 10,000 feet and/or within 30 miles of significant airports to have a transponder. The enhanced coded information on Flight 370's transponder provided controllers with altitude, heading, airspeed, ground speed, roll angle, vertical velocity, and a unique identifier displayed on the controller's screen.

Primary and secondary radar are no longer the only way to monitor flights. With the Global Positioning System (GPS), modern planes can determine their position and speed more accurately than controllers. Controllers are gradually transitioning to the new system known as ADS-B (Automatic Dependent Surveillance–Broadcast). For our purposes, the ADS-B signal can be thought of as a more accurate transponder with transmission occurring nearly continuously. This compares to a traditional transponder waiting to be interrogated with accuracy limited by a receiver dish rotating every 6–12 seconds, with more seconds spent "hunting" for the target and still other technical delays. ADS-B, radar, and the transponder all are limited by line-of-sight transmission to a ground station—about 200 miles. Oceans, the poles, and remote parts of Africa and Asia remain untracked. To keep controllers aware of a flight's position in these remote areas, pilots make regular position reports relaying their position, altitude, and estimated time at the next position on their flight plan.

ADS-B is a critical part of the FAA's NextGen air traffic control upgrade, which is being phased in between 2015 and 2025. NextGen aims to use automation to provide more accurate information to shorten routes, reduce delays, and increase capacity. The switchover depends on building ground stations (634 new stations in the United States) and equipping planes. Boeing and Airbus have been making ADS-B-equipped planes for some time.

There is limited ADS-B infrastructure installed in Southeast Asia, and the missing plane did have a traditional transponder and ADS-B transmissions. The transponder and ADS-B signal were both lost at 1:12 a.m. by controllers located at Conson Island, off the coast of Vietnam. There are reasons a pilot might shut off the transponder. Because airliners have two transponders, in case of failure one is shut off and a second turned on. Controllers can occasionally be confused by planes close on the ground at busy airports and request a transponder be turned off, another problem solved by more accurate ADS-B.

Flight 370 had a more fundamental communications breakdown because the pilots failed to maintain radio contact. The Boeing 777 has five radios: two high frequency (HF) and three very high frequency (VHF). HF waves, which bounce off the earth's atmosphere, are not limited by line of sight but are subject to atmospheric disturbances. If controllers lose contact, they ask nearby planes to check in with VHF radios. The communication failures occurred as the plane was scheduled to pass from Malaysian controllers to Vietnamese, perhaps an ideal time to sow confusion. Vietnam controllers never made contact, nor did the plane appear on their screens. Intentional human behavior explains all these circumstances. No other reasonable theory has yet emerged.

Why Can't We Track Planes over Oceans?

HF radio waves bounce off the ionosphere, but solar activity, thunderstorms, and other disturbances can interfere. It can take 20-45 minutes to send a garbled message and receive a garbled reply. For that reason, plane spacing over the North Atlantic was 120 nautical miles longitudinally and 100 miles laterally. Spacing was reduced with the introduction of the Future Air Navigation System (FANS) in the 1990s. FANS combined GPS location (commercially available in 1996) and a HF data link. Today, longitudinal separation for plane pairs, if both are equipped with the latest FANS, is 30 nautical miles. More direct routes save fuel with FANS, and more than three times as many planes can be safely accommodated. FANS is now mandated on the most advantageous routes and altitudes (35,000 to 39,000 feet) in the North Atlantic, which is the world's most traveled ocean air route.

An estimated 80% of wide-body long-haul planes have FANS. Not every airline activates the system (there is a monthly fee to connect to the network), and not all air traffic controllers are configured for FANS.

ACARS: A Partial Solution

Historically, flight crew pay was based on hours worked between gate times —departing and arriving. The time was logged and radioed to the airline's dispatcher. Because airlines (and federal regulators) are interested in delays and canceled flights, the information became important operational data. The same information is still logged today for the same reasons. Affecting paychecks, airlines were often skeptical of the official log. At that time, large commercial planes (designed through the 1950s) required a flight crew of five: two pilots, a flight engineer (monitored engines and other systems), navigator, and radio operator. Today's jets, with better avionics and communications, need only two pilots.

Aircraft Communications Addressing and Reporting System (ACARS) was introduced by ARINC in 1978 as an electronic time stamp. (Aeronautical Radio Incorporated, later renamed ARINC, began providing aviation radio services in 1929. The ARINC name survives today as a wholly owned subsidiary of Rockwell Collins.)

To track takeoff, landing, and time in and out of the gate, electronic sensors monitored plane components (i.e., parking brakes, landing gear, etc.) and sent out information via telex (telex is a 1930s technology that connects a printer to a radio or telephone). ACARS has evolved into a sophisticated data link communication system. In the mid 1990s, ACARS began sending position data. Today, ACARS provides important data to air traffic controllers, airline dispatchers, and airline mechanics.

ACARS stopped transmitting on Flight MH370 at 1:07 a.m. The Malaysian flight reportedly had a minimal ACARS system that sent out messages only once an hour to the engine manufacturer; the official Malaysia government report is ambiguous on this point. What's important is that the information part of ACARS was somehow disabled on the Malaysia flight. ACARS can be disabled on most planes by accessing the correct computer screen after selecting a sequence of cockpit switches—something pilots know how to do and something potentially discoverable by others. To shut off the communication part of ACARS, someone must enter the electronics bay beneath the cockpit.

ACARS continued to "ping" the satellite system roughly every hour, with the last ping occurring at 8:11 a.m. During an automatic ping (or "handshake"), the satellite essentially asks the plane, "are you still there?" The plane responds, "yes I am." This alerts the satellite network to continue allocating computer resources to listen for Flight 370, although the pings are

Mechanics and the Aircraft Communications Addressing and Reporting System

For safety and redundancy, a large commercial jet has three or four electrical and hydraulic systems. Only one is needed for safe flight. However, it is illegal to take off with a faulty or degraded system. Modern planes have thousands of sensors connected through a local computer network. The computer monitors the data and selectively transmits messages that are eventually routed to mechanics. This information helps mechanics locate spare parts, schedule required work, and avoid costly delays.

not designed to locate the aircraft. Mathematical analysis by the satellite provider yielded vague hints about the plane's location.

Over land, ACARS communicates with line-of-sight very high-frequency (VHF) ground stations. When out of range, ACARS automatically switches to high-frequency (HF) data link ground stations or a satellite network. HF data link is a digitized HF signal. An analog radio wave is digitized by converting to 1's and 0's: on or off—voltage or no voltage. A "1" is a range of voltages, typically 2 to 5 volts. This allows some interference without corrupting the signal. Additional computer logic allows for error checking and retransmission: the computer automatically searches for the best transmission frequency. During a solar storm in 2003, all analog HF voice communication over the poles was lost. Communication was maintained with HF data link. Using just one-third to one-half the bandwidth of analog voice, virtually the entire globe is accessible with just 15 HF data link ground stations.

ACARS messages can also travel on one of two competing satellite networks: Iridium, using low earth orbit (485 miles high), and Inmarsat satellites orbiting about 22,000 miles up. The higher Inmarsat satellites rotate with the earth, remaining stationary—planes need a more powerful transmitter to use Inmarsat. Iridium satellites, orbiting every 100 minutes, require a more complicated tracking algorithm but a less powerful transmitter. Inmarsat is not available over the poles, and because the two satellite systems are incompatible, each plane is equipped to use one or the other, but not both.

Like a consumer paying for cell phone or Internet service, ACARS needs a monthly connection fee. Personnel manage and maintain the network, switching communication from station to station (or satellite to satellite), while computers automatically reroute the data in case of equipment failure and route the information to the air traffic controllers and the airline's dispatchers. Just like in 1929, ARINC remains the biggest service provider, delivering more than 60 million messages per month to over 15,000 aircraft. The smaller (but rapidly expanding) HF data link system supports over 2,700 planes with 1.2 million messages per month.

In addition to ordinary voice high frequency (HF), planes with the Future Air Navigation System (FANS) also communicate with controllers through digital text messages. FANS automatically reports location to air traffic control with Automatic Dependent Surveillance–Contract (ADS-C), not to be confused with ADS-B. The text messages and ADS-C data are sent through ACARS.

ADS-C is both a standard and an application. Except for the satellite link, ADS-C uses the same equipment as the land-based ADS-B. Recall that ADS-B broadcasts every second. ADS-C, working with the limited bandwidth of existing (and somewhat dated) satellite technology, transmits only every 10 to 30 minutes.

Malaysia 370 had ADS-C but was flying within range of ground controllers and not using it. Flight 370 also had ADS-B, high-frequency (HF), and very high-frequency (VHF) radios; a traditional transponder; and ACARS communicating through the Inmarsat satellite network. Its ACARS communicated by VHF radio over land and with satellite communications when out of range for voice and text information, and the flight had inflight entertainment connectivity for the passengers. But none of this matters if it's shut off or not used.

Newer Planes Send Out More Data

Each new generation of plane has more software, computers, and sensors. The Boeing 787 (first flight in 2011) produces 30 to 45 megabytes of data on a typical nine-hour flight. All 787s in the world are monitored in real time (via ACARS) at a central Boeing location in Seattle. The system will even send messages to the cell phones of Boeing engineers. The intent is to proactively respond to any sensor or software glitches. The concern is operational efficiency, not safety. (Because of system redundancies, glitches are not a flight safety threat. To maintain redundancy, the defects must be repaired.) Boeing added position data to many of the ACARS messages. If certain flight safety parameters are exceeded (indicating a potential inflight emergency), position data are updated every 10 or 20 seconds.

Space-Based ADS-B

Tracking planes varies from country to country and airline to airline. Some airlines have complex computerized dispatching systems linked to tracking (to minimize delays and optimize the business of aviation, not for safety);

other airlines do not track their planes over oceans. These sorts of differences make international agreements problematic. For example, the International Civil Aviation Organisation (ICAO is a United Nations agency that holds no legal authority) recommended position reports every 15 minutes by end of 2016 and 1-minute reporting during (unspecified) abnormal flight conditions by 2021. Concerns about airline-to-airline differences in tracking and equipment led to the nonbinding 2016 recommendation quickly being delayed to 2018. Also, there are virtually no plans addressing tamper-proof communications systems, a key issue in the Malaysia 370 investigation. Because of potential fires, pilots are adamant that all electrical systems must have at least circuit breakers.

In the developing global plan, ground-based ADS-B has already become the standard in a few countries and will be mandated on all European planes by 2017 and all US planes by 2020. ADS-B ground-tracking stations, which are less expensive to install and operate, are now the installation of choice worldwide. Two companies are planning on tracking the ADS-B signal with an updated satellite network. This plan requires no changes with existing ADS-B-equipped planes. Iridium plans to provide worldwide coverage with 81 satellites (including 6 orbiting spares and 9 spares on the ground). The spaced-based ADS-B plan got a boost in November 2015 at the World Radio-communication Conference. The frequency band 1087.7–1092.3 megahertz, already assigned for ADS-B ground station transmissions, was assigned for signals to and from space. After repeated delays since 2014, SpaceX finally launched the first 10 satellites for the new Iridium system on January 14, 2017.

Air France Flight 447

Air France Flight 447, lost over the Atlantic in 2009 with 228 passenger and crew, had been the "greatest aviation mystery of the century" before the loss of Malaysia Airlines Flight 370. Flight 447 illustrates the confusion associated with a missing international flight and the difficulty of an ocean search. Despite many advantages not available for the missing Malaysia plane, the search for AF447 still took nearly two years and almost failed.

On May 31, 2009, Air France Flight 447 departed Rio de Janeiro at 7:29 p.m. local time bound for Paris. The Airbus A330, like all modern planes, requires two pilots. This 10-hour flight had a third relief pilot—a second First Officer.

About 3.5 hours into the flight, the 58-year-old Captain (with nearly 11,000 hours of flight time including 1,747 hours in the A330) left the flight deck for a mandated rest break in a small cabin behind the cockpit. The plane was being flown by the First Officer in the right seat (age 32, nearly 3,000 flight hours including 807 in the A330), and the First Officer in the left seat (age 37, over 6,500 flight hours including 4,479 in the A330) was the Pilot Monitoring.

About 8 minutes after the Captain left, the crew was alerted by an audio warning—the autopilot and autothrust had disconnected. This event was officially recorded at 2 hours 10 minutes and 4.6 seconds Coordinated Universal Time (shortened to 2 h 10 min 4.6). The investigation eventually concluded that the pitot tubes temporarily froze, unfroze, and refroze. The pitot tubes of an aircraft measure the force of the air coming into them and send that information to air data computers, which then generate airspeed indications from the data. If the pitot tubes freeze over, accurate airspeed data are lost, and without valid airspeed, a great deal of aircraft automation becomes inoperative, requiring the human pilots to take over flying manually.

The stall warning—which on the Airbus consists of the spoken word "stall" and a cricket chirp sound—briefly went off twice. The Pilot Flying (PF) repeatedly overcontrolled the plane by rapidly rolling excessively right and left. The Pilot Monitoring (PM) stated that the plane was climbing and asked the PF several times to descend. About 45 seconds into the event, the PM began chiming the Captain. The apparently napping Captain returned to the cockpit in 52 seconds.

At 2 h 10 min 51, the thrust levers were changed to takeoff/go around (TOGA, meaning maximum available power). The stall warning triggered continuously for 54 seconds and stopped momentarily before sounding again. With the plane's angle of attack exceeding 40°, the Captain reentered the cockpit at 2 h 11 min 42. The plane, initially at 35,000 feet, was then sinking 10,000 feet per minute. The arriving Captain asked, "What are you [doing]?" The PM replied, "I don't know what's happening."

Vibration noises were recorded in the cockpit from 2 h 10 min 54 until 2 h 12 min 57; the noise was associated with buffeting and aerodynamic stall.

Adding to the confusion, the airspeeds were alternating between reliable and not, while the stall alarm continued intermittently. At 2 h 12 min 02, the thrust levers were reduced to idle and the speedbrake lever was deployed. The pilot tried to slow the plane, apparently believing high-speed buffet was occurring, while in actuality the buffeting was due to stalling. After failing to arrest the dive, the thrust settings were returned to the TOGA setting.

The plane was descending extremely rapidly in a fully stalled state, yet the pilots attempted to pull the nose up, which is exactly the incorrect response to a stall. The angle of attack, when valid (the angle of attack is invalid and the stall warning is inoperative when the measured airspeed is less than 60 knots), always remained above 35°—a fully developed stall. Beginning at 2 h 14 min 17, the verbal alarms "pull up, pull up" and "sink rate" were triggered by the ground proximity warning system. About 7 seconds later, a pilot shouted, "We're going to crash; this can't be happening!" The recordings stopped at 2 h 14 min 28.4: the entire event was over in just 4 minutes and 24 seconds.

Air France 447, compared to Malaysia 370, had an upgraded (and more expensive) Aircraft Communications Addressing and Reporting System (ACARS) service that was not shut off and automatically sent position reports every 10 minutes. After the last ACARS position report, 24 additional maintenance messages (without position) were sent approximately every 5–6 seconds. These messages helped identify the time of crash and help narrow the search to an area of approximately 6,600 square miles. For the Malaysia plane, a team of international experts have somewhat uncertainly prioritized 23,000 square miles to search.

Air France 447's last radio contact occurred at 10:35 p.m. Just 30 seconds later, the Brazilian controller asked Flight 447 three times, without receiving a response, for an estimated time to waypoint TASIL. Because radio communication problems are not unheard of in this obscure corner of the world, the controller lost interest; besides, the plane was scheduled to fly out of his zone of control. Flight 447 routinely flew off of Brazilian radar screens at 10:49 p.m. and failed to report its scheduled arrival at waypoint TASIL around 11:20 p.m. West of TASIL, airplanes report to Brazilian air traffic controllers, and on the other side, Senegal. The plane crashed at 11:14 p.m.

There are internationally agreed-upon "phases" of an aviation emergency at sea. The "uncertainty" phase is triggered if a plane is 30 minutes late communicating. "Uncertainty" graduates to "alert" after repeated attempts at radio contact fail or the plane is 60 minutes late. But somebody must initiate the process and declare an emergency. Nothing was going to save this doomed plane. But delays (and ocean drift) most likely did affect the torturous search for the wreckage and critical black boxes (military planes can search at night, albeit with less precision).

The "alert" advances to "distress" after widespread attempts at communication fail or if a forced landing is expected. Having no experience with large commercial planes that disappear (small planes are another matter), air traffic controllers on both sides of the Atlantic were slow to understand the disaster.

At 11:48 p.m., the Senegalese controller told his Brazilian counterpart there had been no contact with Flight 447. At 12:54 a.m., controllers at Cape Verde and Senegal discussed the location of Flight 447, thinking that perhaps the estimated time of passage was in error. At 1:18 a.m., Senegalese air traffic controllers asked a second nearby Air France plane to contact Flight 447. Failing the attempt, the Air France Operations Control Center was asked to try. By 2:30 a.m., the Operations Control Center was beginning to understand the significance of the automatic ACARS messages sent from Flight 447's diagnostic system. About 20 minutes later, Air France notified French search and rescue (SAR) authorities. French authorities assumed those closer to the scene would respond. Between 11:47 p.m. and 2:30 a.m., there were a series of phone conversations between controllers in Brazil, Senegal, Cape Verde, and the Canary Islands. Despite frequent checking and rechecking, nobody triggered an emergency response.

At 2:23 a.m., Brazilian controllers registered the disappearance of Flight 447, but this action resulted only in preliminary information gathering. Between 2:30 a.m. and 4:30 a.m., there was another flurry of phone calls between Air France; controllers in the Azores, Barcelona, Bordeaux, Brest, Lisbon, and Madrid; Shanwick Oceanic Control; and SAR authorities in France.

At 5:15 a.m., about 6 hours after the last expected contact with Flight 447, the "uncertainty" and "alert" phases were finally triggered by air traffic controllers in Spain. French controllers issued a "distress" at 6:09 a.m. on June 1, 2009. Search planes were eventually launched from Brazil at 8:04 a.m. and Senegal at 9:14 a.m., about 9 and 10 hours, respectively, after Flight 447 failed to make radio contact.

At the beginning of the search, there was still confusion about the last known position (LKP) of the plane. The Brazilians were acting on the position as the plane left Brazilian radar; the Senegalese were using a virtual flight path in their controller's computer. The correct LKP was sent by the Aircraft Communications Addressing and Reporting System (ACARS) to the Air France Operations Control Center. (The missing Malaysian plane is believed to have flown for hours past the LKP.) The Brazilian authorities finally received the correct LKP at 10:00 a.m. An hour later, authorities settled on how the search would be conducted and who would be involved.

With no wreckage or black boxes, 24 ACARS messages sent by equipment diagnostic software between 11:10 p.m. and 11:14 p.m. on May 31 became the focus of the investigation. Early speculation centered on nearby massive thunderstorms, some as high as 50,000 feet. Other planes flying on similar routes reportedly detoured up to 90 miles around storms. Although dozens of small planes are lost every year flying into thunderstorms, no commercial jet in the modern era has been structurally destroyed by a thunderstorm, although they have contributed to pilot errors and accidents. Inexperienced pilots of small planes sometimes become lost and wander into storms. Unlike most small planes, modern commercial planes fly with onboard weather radar.

Far more common is lightning strike, another early suggested cause of the disaster. A large commercial jet is struck by lightning (creating a massive current flow through the plane) about once per year on average. So long as there is no electrical arcing in the fuel tank vapor space by way of burn through or excess current flowing through a gap (keep all those rivets tight), a lightning strike is benign. No US passenger plane has been destroyed by lightning since 1963.

"New Signs That Air France Jet Broke Up in Flight" read the *New York Times* headline 10 days after the crash. More definitive was the *London Sunday Times*: "Crash Jet 'Split in Two at High Altitude.'" Quoting an unnamed Air France pilot, "It's Highly Likely a Bomb Went Off," said the *London Daily*

Mirror. Adding credence to the story was confirmation of a bomb threat on an Air France flight four days before the accident. But all government agencies downplayed terrorism as a cause.

There is a short list of events that have historically broken up planes at altitude fast enough to prevent a Mayday call, including: terrorist bomb, fuel tank explosion, explosive decompression, and inertial loads. In-flight breakup is discussed at length in *Beyond the Black Box.* Briefly, just like a weight on the end of a string will break the string if swung fast enough, a plane in an uncontrolled dive can rip itself apart with inertial loads. (A passenger jet is not a dive bomber!) Metal fatigue, which can be explained by bending a paper clip back and forth until it breaks, can also destroy a plane in flight. If the fatigue crack reaches a critical size, the internal energy of the pressurized air is catastrophically released in an explosive decompression, just like sticking a pin in a balloon. Today, metal fatigue is a solved engineering problem. A serious maintenance error can still fragment a plane, but it's extremely rare, with just one example in the last 26 years. In 1996, TWA Flight 800 famously broke up over the Atlantic after a fuel tank explosion, which was caused by arcing associated with aging wire on the 25-year-old plane. For our purposes, metal fatigue and aging wire are associated with planes much older than Flight 447's 4-year-old Airbus A330.

Citing unnamed investigators, the *Los Angeles Times, Boston Globe,* and *Chicago Tribune* reported that unclothed bodies had been recovered, providing an important forensic clue. The terminal velocity of a spread-eagle sky diver is about 120 mph. A tumbling body from altitude will fall many times faster at a speed expected to rip off most or all clothing. If in fact naked bodies are found, it's virtually certain the plane broke up at altitude, and the list of possible causes is significantly reduced.

Autopsies are an important part of the forensic investigation. Do the bodies have shards of metal, indicating a bomb blast? Do the passengers have smoke or water in their lungs? Does the type of injury correlate with the plane's impact? Later it was announced that many victims had spine fractures, consistent with violent downward impact.

The debris field of an airplane striking the earth is thousands of feet long. If a plane breaks up at 35,000 feet, the debris field can be many tens of miles. Beginning on June 6, the first debris and bodies from Air France

Flight 447 were found about 43 miles from the last known position. Fifty bodies were eventually recovered between June 6 and 17 and, adding credence to the theory of inflight breakup, were scattered over perhaps 170 miles. On July 2, 2009, French authorities announced that the plane struck the water intact. All the bodies recovered turned out to be fully clothed. The bodies were scattered by ocean currents and wind.

This compares to a barnacle-encrusted flaperon (a type of flight control that functions both as an aileron and flap) belonging to the missing Malaysia plane found 16 months after the disappearance. The lightly damaged flaperon resulted in media speculation the jet may have entered the water gently, possibly gliding a significant distance after the engines flamed out. Searchers, groping to identify a search area, had assumed the Malaysian flight entered a steep dive after fuel exhaustion. Assuming a 100-mile glide potentially increases the search area by a factor of 10. As of April 2016, there has been no official announcement about the condition of the flaperon, and no conclusion reached on how the plane entered the water.

Air France 447: The Search

Brazil led the surface search with 20 military and civilian boats and ships, 6 planes, and 3 helicopters. France added 5 ships, a submarine, 3 planes, a helicopter, and the French military Airborne Warning and Control System, commonly known as AWACS. It is a structurally modified Boeing 707 with a rotating 30-foot-diameter radar dome weighing 16,500 lbs. In this case, the AWACS arrived on June 3 and flew 10-hour missions out of Senegal until June 7. It was used to search for floating debris, coordinate collision avoidance among the 25 ships and boats, and relay radio messages. Surface scanning radar covered a radius of 220 nautical miles. Objects identified by AWACS radar might be floating debris, schools of dolphins, or even breaking waves. The AWACS defined targets of interest, requiring visual inspection by a ship or helicopter.

Only about 3% of the plane (bits and pieces of plastics that will float) were recovered on the surface. The largest part, the composite vertical stabilizer about 25 feet long, was found floating flat in the water, and despite its size it was difficult to see. The plane was severely fragmented on impact (Flight 447's last recorded sink rate was 180 ft/sec). Reverse drift simulations attempted to predict point of impact. After just five days of simulated drift, the computer models differed by 60 miles.

The batteries inside the black box pinger die in 30 days or so. During the investigation, French authorities recommended increasing the pinger battery life to 90 days. The mandate finally took effect in the United States and Europe in December 2015, but not in Malaysia or the rest of the world. The simplest changes are more complicated than they appear.[2]

After the black box pinger batteries expired, there were three distinct phases of underwater searches. Each phase was defined by careful study to define a new underwater search strategy. Experts studied (and restudied) ocean drift and Aircraft Communications Addressing and Reporting System (ACARS) messages, trying to focus the search. Despite narrowing it down to a circular area with a radius of 40 nautical miles, the wreckage was not discovered until April 2011, almost two years after the crash. It was rumored the search was close to being canceled.

Malaysia Airlines Flight 370 Ocean Search

Reportedly, almost 60 ships and 50 planes from 26 countries searched for the missing Malaysian plane in a poorly defined, and constantly changing, search area. Like with Air France 447, there were several early and exaggerated reports of floating ocean garbage. By June 2014, the search area focus had shifted for the third time.

After a two-month hiatus to study the potential flight path, a more in-depth search began in June 2014. Bathymetric sonar mapping of over 77,000 square miles of ocean floor was conducted with a surface ship supplied by the Dutch company Fugro. (The Dutch, with a seagoing tradition going back hundreds of years, have many companies that provide specialized world-class ocean services.) Beginning in October 2014, two ships from Fugro began searching with a towed sonar array. With water about 16,000 feet deep, the sonar "fish" was towed on a cable nearly twice as long at a depth of about 500 feet above the ocean floor. The fish will dive deeper to explore items of interest.

The sonar array cannot be towed through extremely rugged terrain. Instead, an autonomous submersible—essentially a robotic or drone submarine—is used. The drone submarine, carrying its own power, is retrieved every 20 hours or so and can only search about 15 square miles on a daily cycle. There are scanning tradeoffs. Higher-frequency sound waves create images with higher resolution, permitting scanning from a greater height with a larger field of view, but higher-frequency sound requires more power that depletes the batteries.

About 1,000 miles from Perth, Australia, and out of range of helicopters, everything including illness requires a one-way six-day trip to Australia. Making the search even more arduous, the search is often suspended during stormy weather. After searching 46,000 square miles, an area a little larger than the state of Ohio, and spending $150 million, the Australian and Malaysian governments announced a suspension of the search on January 17, 2017.

Air France Flight 447 and Aerodynamic Stall

Less lift is produced with reduced speed. To compensate and maintain level flight, the pilot must increase the angle of attack (AOA). At some critical point, increased AOA no longer adequately compensates, and lift rapidly decreases (at a point called the critical angle of attack). This condition is known as aerodynamic stall. Stall is preceded by low-speed buffet caused by flow separation.

As described in Chapter 1, it takes a force to bend the airflow downward along the wing's contour. Lift was described as the equal and opposite force the air exerts on the wing. Lift can also be described as a lower pressure on the top of the wing; as the AOA increases, the pressure on the top of the wing decreases. One simple explanation of this lower pressure: as the AOA increases, the air must flow with increasing curvature as it flows up, over, and around the leading edge, as shown in Figure 4.2.

Air pressure can be considered to be the impact of individual air molecules on a surface. As the AOA increases, the radius of curvature for the air flowing over the leading edge becomes smaller. The air molecules' inertia, which tries to maintain a straight-line motion, flings the molecules outward—the so-called centrifugal force.[3] It's like a car driving around a curve; a tighter curve flings the passengers harder against the door. This outward motion of the air reduces the number of impacts on the wing's surface and thereby lowers the pressure to something less than ambient air pressure.

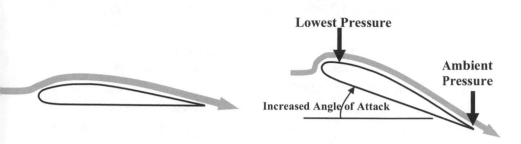

Figure 4.2. Increased angle of attack increases the circular motion of airflow up and over the leading edge. The lowest pressure is somewhere near the leading edge, as shown.

The lowest air pressure occurs somewhere on top near the leading edge. Moving toward the trailing edge, the pressure increases as it returns to normal ambient pressure. Most of the air on top of the wing flows into increasing pressure that resists the air flow. As the AOA increases, the lowest pressure decreases and the pressure differential resisting the flow increases. When the pressure differential increases beyond some critical value, the airflow is disrupted; flow separation (an official engineering phrase) occurs and the wing is in an aerodynamic stall—a dangerous condition.

The flow separation mechanism requires an understanding of the boundary layer, the layer of airflow affected by the presence of a boundary. Surprisingly, the air sticks to the wing's surface. It's difficult to understand how air can stick to anything, but the earth's rotation illustrates the point. At the equator, the circumference of the earth is roughly 24,000 miles. The earth rotates 24,000 miles every 24 hours, for a speed of about 1,000 mph. Yet there is no 1,000 mph wind at the equator because the air sticks to the surface of the earth and rotates with it. The earth's atmosphere is trying to reach 1,000 mph, but the effect breaks down in the turbulence of weather and the thinning of the atmosphere at increased altitudes. (As structures become taller, building codes require higher wind loads; jet streams of nearly 250 mph have occurred.)

On the wing, the increasing pressure toward the trailing edge eventually stops and reverses the direction of the flow, as shown in Figure 4.3.

The smooth downward flow over the wing is disrupted, and lift is greatly reduced.

Flow separation occurs at aerodynamic stall. The amount of air being deflected downward and creating lift is significantly reduced. Flow separation also results in turbulent buffeting. The air is no longer flowing smoothly along the wing but is flowing off in turbulent eddies. The turbulence is violent enough to shake the tail and vibrate the entire plane, as recorded on the cockpit voice recorder of Air France 447 beginning at 2 h 10 min 54. Turbulent buffeting precedes the rapid drop in lift and warns the pilot of impending stall.

Fundamental to the fate of Air France 447, the angle of attack (AOA) is relative to the direction of flight. (To be more specific, the AOA is the angle formed between the chord line of the wing and the relative wind. The critical AOA varies depending on several variables, but in most aircraft it is around 15°.) If the plane is rapidly descending (which it will be in a fully developed stall), the AOA is in fact much higher, as shown in Figure 4.4. The correct response, in defiance of all common sense, is to lower the nose to reduce the AOA, even if the aircraft is descending rapidly. Fortunately, what defies common sense is well known to experienced pilots. In the Air France

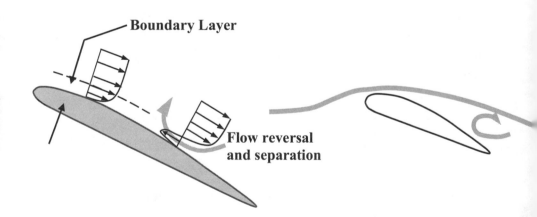

Figure 4.3. An actual boundary layer drawn to scale would be too thin to see or draw. If the pressure differential is large enough, the flow reverses in the boundary layer, resulting in flow separation.

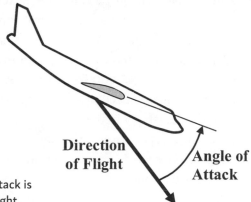

Direction of Flight

Angle of Attack

Figure 4.4. The angle of attack is relative to the direction of flight.

447 accident, three experienced pilots apparently never considered a stall (despite the intermittent stall warning, airframe buffeting, and enormous rate of descent), or at least they never spoke about it as they struggled to understand the problem—they were stuck thinking in the wrong direction. Aerodynamic stall should have been the first thing considered because it explained all the symptoms. It's unclear why the pilots failed to recognize a stall. Captain Hedges adds: One theory is that the on again, off again stall alarm made it appear spurious, though even an intermittent stall warning should have alerted experienced airmen to consider the possibility they were stalling. In retrospect, it's easy to critique the crew's performance, but remember that the crew was sitting in a noisy, buffeting, wildly gyrating machine with numerous cautions and warnings going on, none of which agreed with their (flawed) mental model of the situation. They seemed to believe that they were experiencing high-speed buffet, when in fact they were experiencing stall buffet.

High-Speed Buffet

A plane moving through air creates a series of pressure pulses similar to the waves created by a ship moving through water. As the plane's speed increases, the pressure pulses are forced together or compressed. If the plane exceeds the speed of sound, about 660 knots at sea level, a shockwave or sonic boom is formed. The plane is moving faster than the air can get out of the way; the air compresses and snaps back after the plane passes. (The crack of a whip and a supersonic bullet are examples of a miniature sonic boom.) The speed of sound decreases with temperature, and therefore decreases to 574 knots at 40,000 feet. Engineers identify the ratio of the plane's speed to the speed of sound at a given altitude as the Mach number.

Large commercial jets fly in the range of 0.8 to 0.86 Mach. Because the air passes over the wing faster than the plane's speed, a shockwave forms on the wing at speeds close to the speed of sound. Behind the shockwave there is a rapid and drastic transition from supersonic to subsonic flow, creating an almost instantaneous increase in pressure, temperature, and density and a rapid decrease of speed. As the shockwave strengthens, flow separation (i.e., high-speed buffet) will occur. High-speed buffet will cause an incremental drop in lift and a large increase in drag. The buffeting air may not properly flow across the elevators. And the elevators themselves can form high-speed buffet, resulting in potential control problems. For this reason, a maximum Mach number is established for every aircraft and should never be exceeded.

At one point the Air France Flight 447 pilot reduced engine thrust to idle and activated the speedbrake lever, apparently thinking the plane was experiencing high-speed buffet. High-speed buffet, however, should not occur on the A330; it didn't even occur during certification test flights. Other design requirements[4] limit the plane's speed before the high-speed buffet condition is reached. This extremely low expectation of high-speed buffet was not adequately explained during pilot training. In fact, the exact opposite occurred: the potential for high-speed buffet was given the same prominence as low-speed buffet.

What Is a Pitot Tube, and Why Did It Freeze?

To keep the plane from going too fast or too slow, the pilot (or autopilot) must constantly monitor speed. Recall from Chapter 1 that lift equals ½ × (coefficient of lift) × (air density) × (area of wings) × airspeed^2. Because air at 38,000 feet is only 27% as dense as air at sea level, a plane flying in still air at 38,000 feet will produce only 27% as much lift as the same plane flying at the same speed at sea level. To keep from stalling, airplanes must fly at increasingly higher speeds to compensate for decreased density at higher altitudes. For example, to create the same lift in still air as 130 knots at sea level, the plane must fly at 250 knots at 38,000 feet.

The speed displayed on a pilot's airspeed indicator is the indicated airspeed (IAS). Pilots base their aircraft maneuvers and speeds on IAS. Calibrated airspeed (CAS) is IAS corrected for instrumentation error (this is normally trivial). True airspeed (TAS) is the speed at which the aircraft is traveling through the air mass. Because atmospheric pressure decreases with altitude, TAS increases as a plane climbs at a constant IAS. The ground speed (GS) is the speed of the plane across the ground and is affected by headwinds and tailwinds. A plane with a TAS of 250 knots and a 100-knot tailwind will have a GS of 350 knots, while a plane with a TAS of 250 knots and a 100-knot headwind will have a GS of 150 knots. In either of those cases, the IAS, CAS, and TAS will be the same.

The pitot tube measures ram or impact pressure,[5] similar to what is felt by sticking your hand out the window of a moving car. (The force to hold a 1-foot-square plate out the window of a car moving 60 mph is about 10.5 lbs, a pressure of 10.5 lbs/144 inch^2 = 0.073 lbs/inch^2 or psi.) Considering pressure as impact of individual air molecules, the impact pressure increases with increasing wind blast speed. The pitot tube measures the kinetic energy per unit volume of the moving air equal to ½ × (air density) × airspeed^2. This value changes in direct proportion to the lift force and automatically adjusts for changing density and changing wind speeds. The ram pressure measured by the pitot tube is electronically converted into what is called IAS by pilots.

Pilots only care about indicated airspeed as measured with the pitot tube. If a plane is flying with a ground speed of 300 knots and a tailwind of 170 knots, the true airspeed will be 130 knots. The calibrated airspeed for a plane flying at 250 knots in still air at 38,000 feet remains 130 knots, because the ram pressure reflects the reduction of air density at altitude. The pilot only has to monitor the indicated airspeed and can ignore effects of altitude and wind speed. Unfortunately for Air France Flight 447, the pitot tubes froze and unfroze repeatedly.

Can't they prevent pitot tube icing with more heat? It takes large amounts of heat to maintain temperature when the surface is moving many hundreds of knots in –50°F air. There are limits on how much heat the heating element can create; otherwise, it will overheat in a milder environment and potentially burn out or reduce reliability. Bigger heating elements can solve the problem. Presumably, the airplane manufacturer must micromanage all power and external drag and specifies limits on both to the pitot tube designer.

The real problem is supercooled water droplets, or water with a temperature below the freezing temperature. How can that occur? One partial explanation is that surface tension maintains the droplet in a spherical shape that resists ice formation. Touching any surface disrupts the sphere, which results in immediate freezing. Other variables are droplet size (drizzle droplets present unique problems) and distribution, total water content, how long the planes flies through icing conditions, and finally temperature and air pressure. It's even more complicated because there are different crystalline forms of snow and ice. Historically, there have been too many variables to identify worst-case scenarios during wind tunnel testing or even during test flights. (Icing also affects airplane wings and engines; it is a vast topic worthy of its own chapter.) This problem has resulted in significant new research that has led to new Federal Aviation Administration icing rules, introduced in 2015.

Captain Hedges Explains: Air France Flight 447

Fly-By-Wire

An understanding of fly-by-wire (FBW) technology is fundamental to understanding this accident. For decades, aircraft had flight control systems that could be both mechanically extremely complicated and relatively heavy. In the cockpit, pilots moved control yokes (or in some smaller and military aircraft control sticks) to make the aircraft turn, climb, and descend, while their feet used rudder pedals to move the rudder and apply wheel brakes. Additional controls were added for speedbrakes, flaps, slats, trim, and so on, all of which added to the complexity of the flight control systems. All these systems used mechanical controls involving cables, bellcranks, pushrods, jackscrews, and a variety of other complex parts that required both strict manufacturing tolerances and extensive (and costly) routine maintenance. In early or light aircraft, these mechanical controls normally were directly connected to the appropriate control surfaces and were therefore controlled entirely by force directed through the cockpit controls by the pilots. As aircraft got bigger and faster, they required hydraulically powered controls, and the mechanical assemblies controlled by pilots now controlled hydraulic actuators. For safety and reliability, redundancy became the order of the day in airliners, with anywhere between two and four hydraulic systems powering control surfaces. This made the systems in a large aircraft extremely complicated and unwieldly. Many aircraft had additional complexity owing to having provisions allowing for control of aircraft in the event of hydraulic failure ("manual reversion"), though by the end of the 1960s, redundancy was being obtained by additional hydraulic systems in the new generation of wide-body aircraft (B-747, L-1011, and DC-10) because the control surfaces were too large to move manually (most pilots consider a manual reversion approach in the relatively smaller B-727 or B-737 to be an extremely challenging and physically exhausting exercise).

The first digital FBW system without a mechanical backup was designed and operated by the National Aeronautics and Space Administration on an F-8 Crusader in 1972 using a modified Apollo Guidance Computer that was integral to the lunar landings. The military got there next with the F-16, followed by Airbus (though the Concorde had an early analog FBW system years before Airbus).

To reduce pilot error and provide flight envelope protections unattainable with conventional flight control designs, Airbus pioneered FBW technology in commercial aviation, starting with the A320 and all subsequent Airbus planes. So successful was this advance that Boeing also now uses FBW on the B-777 and B-787, although each manufacturer has different flight control philosophies.

The A320 may have looked like a conventional aircraft on the outside, but inside it was as revolutionary as any airliner ever built. Conceived as the first member of a family of aircraft that would have similar cockpits (or "flight decks"), systems, controls, indicators, and flying qualities, the A320 entered service with Air France on April 18, 1988, and has been extremely popular in the marketplace and with pilots. The cockpit of the A320 series and its descendants from the A330 to the A380 share a lot in common, far more than most competing product lines, allowing for much shorter training courses for pilots transitioning from one aircraft to another across the product line. Much of this is attributable to the FBW architecture integral to Airbus products, and the bulk of this section uses Airbus FBW for discussion purposes.

Inherent to FBW are the flight control computers (FCCs) that comprise the system and the cockpit controls that pilots use to interface with them. Airbus chose to forgo traditional yokes and instead use ergonomically optimized side-stick controllers (SSCs) on the outside of each pilot seat (see Figure 4.5). This freed up enormous amounts of space in the cockpit and allowed installation of foldable tables used for charts and the like, but perhaps the biggest advantage of the SSC is that it removed the yoke from its prominent position in front of the instrument panel, as in some aircraft the yoke is an impediment to observing instruments, particularly in the bottom center of the panel in front of the pilots.

There are several types of FCCs in different designs, but their functioning is generally transparent to pilots. They basically serve to take their programming and various inputs about aircraft performance (altitude, temperature, airspeed, angle of attack, etc.) and position hydraulic valves to ensure the resultant flight path is what the pilot desires. The system functions by interpreting external and pilot inputs and quickly solving equations of motion to achieve the desired result. These equations are integral to the aircraft's "control laws." These laws have several variations depending on aircraft status and situation. The A320 has a total of seven FCCs of three different types, and the aircraft can fly with multiple inoperative computers. The redundancy that used to be achieved by increased cabling and mechanical components has now been largely achieved with multiple computers working in unison; even though the exact nomenclature of computers and modes varies slightly between aircraft models, the results are essentially transparent to pilots. FBW requires huge amounts of testing to ensure that all modes result in predictable and desirable control responses in a huge variety of conditions. Both Airbus and Boeing FBW aircraft (Boeing's B-777 and B-787 are sophisticated aircraft, though their control philosophy is much more conventional, with a traditional control yoke, pitch trim that is not entirely automated, etc.) have proven to be incredibly reliable and substantially safer than the previous generation of aircraft.

Figure 4.5. A380 Cockpit. Wikimedia Commons

So how do the control laws work? First, understand that there are several different laws possible, each with its own set of logic and equations to allow the pilot to fly safely in any regime. The A320 control law that is normally in use provides the most assistance and protections for the pilots and unsurprisingly is called "normal law." Remembering that the SSCs have no mechanical connection to flight control surfaces; when a pilot makes a pitch or roll input, it is directed to the FCCs, which interpret the pilot input and direct hydraulic control valves to reposition and move the flight controls to yield the aircraft movement the pilots desire. Because it is so automated, normal law enables several protections that would be impossible without FBW. Among the protections available are high angle of attack protection (a FBW Airbus in normal law can't be stalled, but a FBW Boeing can), load factor protection (the aircraft can't exceed the 2.5 G limitation for transport category aircraft, even if the pilot moves the SSC aggressively), and pitch attitude and bank angle protection, in which the aircraft is limited to a maximum of 30° nose-up and 15° nose-down pitch and 67° of aircraft bank (67° of bank in a level turn requires 2.5 G, the regulatory maximum). Normal law becomes active shortly after takeoff (on the ground, the controls command a response from the control surfaces proportional to the SSC deflection), and any pitch input (forward or aft SSC movement) signals a G load, while neutral SSC commands 1 G. The aircraft uses an autotrim system that keeps the plane constantly in trim regardless of speed or configuration changes (so long as aircraft bank is less than 33°); for roll control deflections, an SSC commands a roll rate and is independent of airspeed. That is, a given movement of the SSC will always command the same roll rate even at vastly different speeds, a trait unlike conventional aircraft. Although some specifics of the Airbus FBW system differ from other aircraft, pilots typically adapt to the natural feel of the system and its functionality rapidly.

If an Airbus aircraft experiences multiple malfunctions, the flight control laws may change to a more basic level of functionality and enter "alternate law" (note that this is a general discussion and that there are slight variations of the actual functionality of alternate law between Airbus models). In alternate law, pitch inputs still command G load, and autotrim is still available, but most protections are lost, allowing the pilot to override certain limits that were strictly enforced by normal law. In alternate law, the aircraft can be stalled; likewise, although it has certain built-in stabilities, it is possible to overspeed the aircraft, or exceed 67° of bank. There are other differences, but these are the key points. The most basic of the control laws is "direct law," which generally occurs after landing gear extension when operating in alternate law. In direct law, there is a direct correlation between SSC deflection and aircraft control surfaces; there are no protections or autotrim available, though pilots can manually trim using controls on the center pedestal.

Other manufacturers have adopted FBW after the successful introduction on the A320, and it has proven to be robust and reliable while reducing weight, maintenance costs, and mechanical complexity of flight control systems, all while improving ergonomy and flight safety. Though some other manufacturers have elected to retain a more conventional cockpit and control architecture (e.g., control yokes and the requirement to manually trim the aircraft), FBW in all its various forms yields a net improvement in safety and economy, and is the way forward in airliner design.

Avoiding Stalls

From early in a pilot's training, there is great emphasis on stall recognition and recovery. An aerodynamic stall occurs when the wing of the aircraft exceeds the critical angle of attack (AOA; also called "Alpha"). It does not occur at a fixed speed, but at a fixed AOA. A plane can be stalled in any attitude (even upside down) so long as the critical AOA is reached. Most pilots first see stalls demonstrated in light single-engine training aircraft. In these type of aircraft, as the plane approaches the stall, there is notable airframe buffet that intensifies as the plane approaches the stall; when the aircraft reaches the stall, the nose will generally fall abruptly (it is common for one wing to stall slightly before the other, and that in combination with a few other factors tends to make one wing drop more than the other). This can be an alarming event, so it is practiced and practiced until the recovery is perfected. The recovery in most aircraft involves adding full power and lowering the nose. When pilots finally get to a more advanced level and begin flying jets, almost all training is done in the simulator, and the vast majority of simulators are only able to accurately replicate the aerodynamic behavior of the aircraft up to the stall.[6] Because pilots operating these aircraft should have great experience with stall recovery fundamentals, most training programs have the pilots recover only from an approach to stall (which is usually indicated by airframe buffet or an artificial warning such as a stick shaker, a device that physically shakes the control yoke or triggers electronic warnings of some variety). Despite this training, aircraft still crash while stalled, and regulators are now issuing more comprehensive regulations that require simulators to accurately model stalls and pilots to demonstrate recovery from fully developed stalls in all airline flying training syllabi.

Although a stall is caused by exceeding the critical AOA whether in a Cessna or an A330, aircraft behavior is different between light general aviation aircraft and airliners. Some airliners exhibit little buffet as they approach the stall; on most aircraft, an approach to stall triggers a stick shaker to provide an unmistakable clue to the pilots that the aircraft is about to stall. The recovery is similar to light aircraft in most large jets: add power and lower the nose, decrease AOA, and recover from the descent. As an aside, some aircraft with large engines under the wing do not immediately use full power because the recovery can be complicated when thrust is suddenly added below the center of gravity, which causes an abrupt pitch up, often resulting in a secondary stall. These are the kind of nuances that airlines train pilots for, specifically when they learn a new aircraft (and are retrained on regularly in recurrent simulators).

In the case of Air France 447, the issue was seemingly recognizing the stall in the first place. After experiencing a temporary loss of airspeed indicators, the aircraft reverted to alternate law (alternate 2B, specifically) and the autopilot and autothrust disconnected, suddenly requiring a pilot to fly the aircraft. At this key juncture, a return to basic flying skills was demanded, where the first order of business must be to maintain aircraft control. In primary training, pilots learn how to fly aircraft with basic pitch and power settings. These same techniques should be taught to pilots when they train on new airliners, as they allow a pilot to stabilize a bad situation and start from a known point. Immediately before the autopilot disconnection, Flight 447 was cruising in level flight at 280 knots indicated airspeed (0.82 Mach) with the pitch 1.8° nose up, and the engines turning at about 99% N1. Appropriate pitch and power settings immediately after the incident would therefore be around the same values. Instead, the Pilot Flying started making rapid, cyclic roll inputs that resulted in both left and right banking and pitched the aircraft up to 11° nose up (ultimately getting even higher), and picking up a 6,700-foot-per-minute climb in the process. This was an unsustainable situation, as a heavy long-range airliner cannot ever climb at 6,700 feet per minute without losing airspeed and increasing AOA, and so it was with Flight 447: as it zoomed from 35,000 to 38,000 feet in less than a minute, the pilot holding the SSC aft, the plane rapidly approached the critical AOA and in less than a minute stalled and reversed from an aggressive climb to a descent of 14,800 feet per minute, a mind-bogglingly huge descent rate that would never be seen in normal operations.

Airbus fly-by-wire (FBW) aircraft do not have stick shakers because the manufacturer and regulators considered them to be unnecessary, for several reasons. The first is that in normal law, Airbus aircraft cannot be stalled, and the overwhelming number of flights never get out of normal law. Second, even if out of normal law, the aircraft still has stabilities in the FBW system that make it easily controlled; even in its most basic form law (direct law), Airbus aircraft are no more challenging to control than other conventional aircraft from other manufacturers. Airbus also installed an aural warning that says "stall" when the aircraft is in a stalling situation: this alert was recorded repeatedly on the cockpit voice recorder, yet the pilots didn't acknowledge it or execute a stall recovery. Possibly compounding their confusion was the aircraft's pitch attitude, which was nose-high for the bulk of the descent, when in most stall situations the crew had practiced earlier in their career, the nose would have likely been below the horizon. Add to the situation the fact that the flight was at night, over water, and near convective weather, and the situational awareness of the crew was totally eroded.

The information was available for the flight crew to solve the problem, yet they didn't. The issue was basic airmanship, which is a key reason that regulators and airlines started training pilots much more in depth about stalls and unusual attitudes and the recoveries from them, both in initial and recurrent training. Additionally, most airlines have begun having their pilots "hand fly" the aircraft (i.e., turn the autopilot off) more in training, with some encouraging it in line flying when conditions permit. This is most critical in automated aircraft like the A330, as the autopilot is on for the majority of the flight time. Many airlines have also been putting extensive instrument failures into training curricula to help pilots make the transition from fully automated flight to manual flight using basic pitch and power techniques. By 2019, new simulator software will be required to model the entire stall regime effectively, and pilots will be required to recover from a variety of stalls on a regular basis. Some onlookers expressed concern that Air France 447 was an Airbus-specific issue arising from the cockpit architecture of the separate side-stick controllers and lack of a stick shaker. In truth, there was more than adequate information available to fly the aircraft and to recover from the stall. Why this highly qualified crew was unable to successfully fly the aircraft will never be definitively known: they were certainly presented with a challenging (but not insurmountable) situation, but if they had been able to transition back to basics more rapidly, the accident could have been avoided. To be clear, stall and loss-of-control accidents are not an Airbus-specific threat, as far more conventional aircraft have stalled and crashed in the history of aviation than Airbus FBW aircraft.

Although the Colgan Air crash in Buffalo, New York, in February 2009, a few months before Air France 447 started the focus on stall awareness and training, there have been other major accidents involving stalls since then. On July 6, 2013, Asiana 214, a Boeing 777, crashed in San Francisco, California, in a stall-related accident in which cockpit automation modes were not correctly understood by the pilots. When the aircraft got low and slow on the approach, they waited too long before initiating a recovery and crashed short of the runway. The aircraft was operating as designed, and despite aural warnings and a stick shaker, the crew did not have adequate situational awareness to prevent the accident. Most recently, on December 28, 2014, Air Asia 8501, an A320 (which has fly-by-wire features similar to the A330) crashed after the First Officer lost control of the aircraft while the Captain was performing an unauthorized procedure (involving pulling circuit breakers) to reset a defective rudder travel limiter. In a set of circumstances similar to those of Air France 447, when the Captain was doing this troubleshooting the autopilot and autothrust disconnected and the aircraft degraded to alternate law, requiring the First Officer (FO) to hand fly the aircraft. As in Air France 447, the FO experienced some roll oscillations (caused in part by the aircraft rudder moving 2° to the left) and pulled the side-stick controller nearly full aft, resulting in an 11,000-foot-per-minute climb and subsequent stall. Also as in Air France 447, the aircraft fell to earth in a descent rate of about 12,000 feet per minute, with the nose approximately level with the horizon. The audible stall warning and stall buffet are clearly audible on the cockpit voice recorder, yet the crew was unable to assimilate this information, identify the stall, and recover from it. The ultimate crash killed all on board.

The ability of pilots to recognize and recover from stalls has been a problem since the dawn of aviation but has recently become a top priority in the industry, as loss of control accidents are the deadliest accidents in commercial aviation, accounting for only 2% of all accidents but yielding 25% of accident fatalities in 2006–2013. Fortunately, regulatory authorities, simulator manufacturers, and airlines are taking this threat seriously, and the increased focus on maintaining basic stick and rudder skills as well as airmanship should lead to a decrease in stall-related accidents.

New Simulator Rules

Pilots fly up to impending stall during training in the simulator and practice recovering from these approaches to stalls. Because the simulator programming does not (yet) have information for how the aircraft behaves beyond the stall, most pilots do not have actual experience recovering from a stall in an airliner.

The whole point of certification test flights is to verify that nothing unexpected occurs at the extremes of the certified (and permissible) flight envelope. In setting boundaries for the in-service aircraft, flight tests intentionally exceed limits placed on airline aircraft to ensure a margin of error in the interest of safety. The test pilots who fly these missions carefully plan their flights with engineers and performance specialists and test-fly the maneuvers in an engineering simulator before ever trying it in the real aircraft. Despite the advances in computer modeling and simulation, real flight tests always seem to uncover a characteristic that is not exactly as predicted. Data from these test flights are used to accurately calibrate simulators, ensuring that the pilots in training will receive a high-fidelity representation of flying the actual aircraft. It can be dangerous to fly in excess of the certified limits, as is evident from the numerous accidents that have historically occurred during training flights. Today, training flights at the operational limits have been completely replaced with better simulators.

The only constant in aviation is change, and simulator training is changing again with new mandated FAA rules. By 2019, all commercial pilots will fly, on simulators, into fully developed stalls and practice recoveries in various scenarios. Boeing, working with the National Aeronautics and Space Administration, has developed new stall models that will eventually be implemented into the simulators. To prove that computer simulation and wind tunnel tests could adequately substitute for test flights, Boeing test pilots repeatedly stalled a 737-800. The two Boeing test pilots had, between them, over 700 stalls.

5

Practice Makes Perfect

The MD-80 droned through the night sky en route from Phoenix to Chicago.[1] It was a classic "all-nighter," leaving Phoenix at 11:56 p.m. with a scheduled arrival time of 5:02 a.m.; it was also the end of the four day long trip for the pilots, and was the third and final flight of their day. Departure had been on time, and the initial departure and climb to the cruising altitude of 33,000 feet had been uneventful, with a smooth ride after passing the Superstition Mountains just to the east of Phoenix. The three Flight Attendants were performing their inflight service, and most of the 140 passengers on board were asleep. In the cockpit, the two pilots were not only awake, but alert and busy looking at the weather they were approaching. With the autopilot on and coupled to the flight management computer (FMC), they were monitoring their progress as they examined the line of thunderstorms just to the east of the Rocky Mountains, largely concentrated around Denver. Although they were still over 100 miles away, the distant flashes of lightning on the horizon were clearly visible, as was the depiction of storm cells on the aircraft radar.

Pilots at most airlines take turns flying the aircraft on alternating flights, and now it was the First Officer's (FO's) turn. Although the Captain is by law ultimately responsible for the safe operation of all flights, FOs at a major airline are experienced and fully licensed to fly the aircraft. For FO currency and to reduce workload on Captains, it's typical to have the FO fly alternating flights (or "legs"). But sometimes experience is hard to precisely quantify. In this instance, the Captain had recently upgraded from being an FO on a larger international aircraft, the B-767. He had never flown the MD-80 before upgrading to Captain, and was in fact still on "high minimums," meaning he couldn't fly to the lowest approach minimums available, as he had less than 300 hours of flight time as an MD-80 Captain. This situation is not uncommon; in this case, the Captain had previously been a B-767 FO and before that a B-737 FO with the company. Although the FO is second in command, on this flight the FO had been flying the MD-80 for over 10 years and had flown the similar DC-9 for years before that. He had about 10,000 hours logged in the type. Although not in command, he was an experienced asset to have on the flight.

In smooth air, the aircraft worked its way across the desert southwest. With no advance notice, the pilots heard a loud bang, similar to a car back-firing, felt a loud jolt, and the aircraft started to vibrate. The weather, which had been their preoccupation moments earlier, was no longer a priority, and the pilots quickly analyzed the situation. The Captain's first action was to ensure the FO was still flying the aircraft, knowing that many aircraft had been lost when the entire crew was working on solving a problem. Things happened in rapid succession: the Flight Attendant call bell went off almost immediately, as the crew tried to focus on the engine indicators to determine the cause of the problem. Several amber lights came on to sig nal the problem, but the primary issue was that the aircraft was vibrating so badly it was difficult to read the digital instruments and the messages. Although something was seriously wrong, the crew also knew that inadvertently shutting down the wrong engine would be a serious mistake that would likely prove fatal, which it had several times in the past. With that in mind, the crew forced themselves to consciously slow down and evaluate the situation before making any hasty decisions. The Flight Attendant call bell went off again. They would have to wait. There is a rigid hierarchy in aviation that is commonly summarized in the three words: "aviate, navigate, communicate." Pilots use that as a mental framework to prioritize workloads. The first element, aviate, means to ensure someone is flying the aircraft properly at all times. It seems basic, but accidents prove otherwise. "Navigate," the second priority, ensures that the aircraft is being flown where it should be going at any given time; a corollary is knowing where the aircraft is in relation to the terrain, the weather, other aircraft, and the airports in the area. Finally, "communicate" comes last. Once the aircraft is being flown safely in the proper direction, it's important to coordinate with all the entities that interface with the operation. In most cases, air traffic control (ATC) gives the aircraft route and altitude clearances, but in case of emergency, the crew may have to act first and then and inform ATC of their actions. In this case, the radio traffic was light, being the middle of the night, and communications with Denver Center were uncomplicated.

As the pilots carefully parsed the myriad instruments in front of them, they both recognized that the issue was with the right engine. While the aircraft experienced several more bangs and jolts, the right engine indications began to roll back, and the vibration lessened. It was becoming clear that they had suffered a failure of the right engine with some kind of engine damage. As part of the "aviate first" philosophy, the FO had selected maximum continuous thrust on the left engine and had begun slowing to the single engine drift down speed. Accessing the speed and altitude target data from the FMC, the FO saw that they would no longer be able to maintain their altitude of 33,000 feet but would have to descend to 25,000 feet owing to the loss of thrust. The plane decelerated slowly, though, and the Captain declared an emergency with Denver Center, and coordinated for an altitude of 25,000 feet as they commenced the Engine Failure–Severe Damage or Separation checklist. The checklist quickly took them through all the steps required to secure the damaged engine and fly the aircraft safely to a landing. The autothrottles were disconnected, the right throttle was retarded to idle, the fuel control lever for the right engine was brought to off, and the right fire handle was pulled (even though there wasn't a fire, this action ensured that the engine systems were all isolated). All these steps were verified before they were accomplished, to prevent accidentally shutting down the good engine. (It's happened before on numerous occasions; refer to the East Midlands 737 accident in *Beyond the Black Box* for a prime example.)

As soon as the checklist had been completed, the Captain finally had enough time to stop and talk to the Flight Attendants, who were unnerved by the noises and relayed that the passengers were scared. The Captain told them that they had experienced an engine failure and would be diverting, but didn't know where yet. He also told them to expect a normal landing but that fire trucks would be visible, and that no evacuation was expected. He told them he'd call them back with more information as soon as he knew. Although it seemed like it had been a long time to the anxious Flight Attendants, less than two minutes had elapsed since the beginning of the incident. As the Captain was speaking to the Flight Attendants, the FO was using the aircraft's datalink (the Aircraft Communications Addressing and Reporting System, or ACARS) to request weather information for the four closest airports that had sufficient runway to land an MD-80. Given the aircraft position, the available airports were Cannon Air Force Base in New Mexico, and those at Denver, Pueblo, and Colorado Springs in Colorado. There were pros and cons to each airport. Federal air regulations (FARs) require that a two-engine aircraft that has an engine failure must land at the nearest suitable airport. But what is suitable? The Captain must consider all available information before making a decision. Cannon was closer by a few miles, but it had no passenger facilities, was used to dealing with small tactical jets, had relatively poor emergency services, and no accommodations for 145 civilians at 1:05 a.m. It was designated as an emergency divert airport in the airline's operations manuals, but preference was always given to civilian airports, preferably with airline service. Pueblo was next closest, but it had the shortest runways. Additionally, the pilots had no instrument approach charts for Pueblo, neither had ever been there, it was night, and there was mountainous terrain to consider. Denver would have normally been the obvious choice: both pilots were familiar with the airport, and it had long runways, good facilities, and was relatively far from the mountains. Unfortunately, it was also being pelted by thunderstorms at that moment. Denver was out. When looking at the weather for Colorado Springs, it was a relatively high overcast sky, about 2,000 feet above the ground, and 3 miles of visibility. The winds were calm, and as the evening progressed and the temperature dropped the probability of fog would increase, but the airport was only about thirty minutes away. Colorado Springs has long runways, airline operations, and excellent emergency response capabilities. Not only that, but both pilots had been there numerous times. After conferring briefly, the Captain asked for an opinion from the FO, and he preferred Colorado Springs. The Captain agreed, and they coordinated for a direct routing to the airport. The Captain then called the Flight Attendants back and told them to expect to land in about thirty minutes in Colorado Springs, and made an announcement to the passengers explaining what had happened and what the plan was. Time elapsed from the start of the incident was about five minutes.

Pilots train intensively to be able to execute demanding procedures in a well-defined sequence of priorities under high stress. As they continued to descend, they had a brief time to regroup and catch up. The Captain asked the FO if he could think of anything they might have forgotten so far, and the FO asked if he had informed the company of the diversion yet. The Captain said no; he then used the datalink to advise the dispatcher about the problem, and that they were diverting to Colorado Springs. Airlines are required to know where their planes are going, and the datalink message satisfied that requirement; it also set into motion a series of steps to get personnel to Colorado Springs to meet the aircraft and to take care of the passengers, crew, and the aircraft. As they approached Colorado Springs, the Captain told air traffic control (ATC) the specific nature of the emergency, the number of people on board, the fuel remaining, and what assistance was required. In this case, he wanted the fire trucks to meet the aircraft and inspect it once stopped on the runway, and if all looked good, to follow the aircraft to the gate until it was parked and the people could be deplaned. The crew received information about the weather and approach in use at Colorado Springs, and it was yet another surprise. They wanted to land on the longest available runway, which in this case was 35 Right (35R), but many airports do maintenance on their navigational aids in the middle of the night because they are often not being used. Unfortunately, the instrument landing system for 35R was out of service, and the only available approach was a much more difficult to fly nondirectional beacon (NDB) approach to runway 35 Left (35L), which would require the pilots to use an old-fashioned needle superimposed on a compass to guide the aircraft. NDB approaches are relatively inaccurate, and after descending through the cloud layer, the pilots would have to sidestep to 35R to get the longer runway, all of this at night, using a single engine, with mountains just to the west of the field.

The Captain had a lot to consider. Should he fly the aircraft, or should he let the FO continue to fly? In the past and at some companies to this day, the culture is that the Captain will fly the aircraft in an emergency. The Captain is generally more experienced and definitely legally responsible for the outcome, but in this case the Captain realized he had been flying the MD-80 for 6 months compared to more than 10 years for the FO. The FO had been doing well throughout the emergency thus far, and with that in mind, the Captain asked the FO how he was feeling and whether he wanted to keep flying. The FO responded that he was fine to fly, and the Captain elected to let the FO continue to fly the aircraft. As the aircraft descended, all the normal checklists had to be accomplished along with other duties. Because the approach was so complex with so many considerations involved, the FO gave the Captain the aircraft control while he studied and briefed the approach. It took them down to 553 feet above the ground using nothing but 1930s technology. The FO briefed what they would do for the landing flap setting and go around. Normally, 40° of flaps would be used to land and 15° would be used in the event of a go-around, but owing to the loss of thrust on one engine, drag had to be reduced. So 28° would be used for landing and 11° for the go-around. Additionally, they coordinated for a different go-around procedure, as the one published exceeded the capabilities of the aircraft on one engine. After a thorough briefing, the FO took back control of the aircraft, and a series of checklists were begun to configure the aircraft for landing. The FO expertly established the aircraft on the 351° course to the Petey NDB, and when he passed it began the descent to the minimum descent altitude. Once out of the clouds, the FO made a maneuver to align with the adjacent longer runway and landed in the touchdown zone on the centerline. After coming to a stop, the Captain made an announcement for the passengers to remain seated (uncommanded evacuations had happened in the past when passengers panicked). After the fire chief looked over the aircraft, he called the crew and told them there was no fire and cleared them to taxi to the gate. The Captain prepared for taxi by looking at the taxiway diagram.

"OK, knock it off. I have the sim." Bright white florescent lights suddenly bathed the cockpit. "Motion's coming down." The crew could feel a settling motion and hear an audible alarm outside as the simulator they were sitting in descended on the six giant hydraulic cylinders that supported it and gave it the motions of flight. The "flight" that had ended in Colorado Springs was in fact entirely simulated in a sophisticated machine packed with computers and perched on top of six hydraulic actuators capable of providing extremely lifelike motion. Everything in the scenario, from the flight planning documents to all the various interactions with air traffic control, Flight Attendants, and the like, were painstakingly reproduced to yield a high-fidelity approximation of the emergency. "How do you think it went?" asked the instructor, sitting at a computer console that ran the simulator just behind the cockpit area. The pilots were hesitant to start critiquing their performance, but once the discussion got underway, there were lessons learned by all three people in the simulator. The crew had done very well. Every six months to one year, airline crews practice and are tested in the simulator. These devices mimic flight so realistically that, in most cases, the first time a pilot lands a real aircraft, he has paying passengers in the back, along with a line check airman accompanying him to get him some experience in his new aircraft. Simulators have progressed rapidly over the years, particularly in the last 20 years as increased computing power has enabled better motion platforms and visual systems so good that in some simulators you can see the grass blow from the correct direction during ground operations if you have the time to look. These simulators cost millions of dollars, frequently more than the aircraft they represent, but they have several huge training advantages. First, they are much less expensive to operate than aircraft, but most importantly, they can simulate emergencies far too risky to practice in actual aircraft. Before certification could be done in simulators, pilots had to go fly the actual aircraft in compromised states, such as with an engine out or with unusual flap settings. Aircraft with three or more engines had to practice two engine-out approaches, and numerous aircraft were lost in training accidents over the years. Now pilots can practice complex scenarios like the en route engine failure between Phoenix and Chicago as well as potentially deadly maneuvers such as escaping from wind shear or following a traffic avoidance command. Airlines still must train certain things by rote—such as different approaches, engine-out procedures, fires, evacuations, and so on—but much more emphasis is now placed on developing good crew coordination in line-oriented evaluations like the above example. The pilots in this scenario were free to pick who flew, what airport to divert to, and how they wanted to divide up cockpit workload so long as the result was safe and procedurally compliant. This kind of simulator training is much more challenging than in the past, but it reinforces good crew resource management (CRM) and has had a very positive impact on flight safety since CRM started in earnest in the 1970s.

Cockpit Tour

To understand exactly what is being simulated, we must look inside the cockpit. An orientation to an airliner cockpit, now commonly referred to as a flight deck, is helpful background information for anyone wishing to understand a pilot's working environment. Many people are surprised at the extreme compactness of most cockpits: these areas are built for crew efficiency and are in many aircraft types cramped, hot, and noisy. Modern aircraft have two seats for the operating crews, plus a minimum of one additional crewmember seat (called the "jumpseat") for purposes of evaluation. In the United States, the Federal Aviation Administration frequently shows up unannounced to perform line evaluations on the pilots to ensure that they are operating in accordance with FAA regulations and company procedures. Company line check airmen also frequent the jumpseat to give no-notice line checkrides to observe pilots. These are all "jeopardy" events, so called because both pilots' certificates are in jeopardy if they make substantive errors. Some aircraft have more than one jumpseat in the cockpit (two or even three are not uncommon), and especially for international flights where more than one pilot crew is on board, other crewmembers will ride in them for takeoff and landing at a minimum. Some older designs have a third required crewmember, a Flight Engineer (or Second Officer) seated behind the pilots on the right side of the aircraft with controls for the aircraft systems in front of him. Newer aircraft have more automated systems and have dispensed with that position. (For an illustration of this advance, compare the cockpit of a 747-200 with that of a 747-400.)

In front of the pilot seats in all older aircraft and in all Boeing aircraft is a control yoke, which is used to steer the aircraft. As in a car, turning the wheel left commands a left turn, while turning it right commands a right turn. Unlike a car, pulling the yoke back toward the pilot pitches the nose up, while pushing forward on it lowers the nose of the aircraft. First-time observers are sometimes surprised at how delicate the required control movements are, especially at high speeds. In some aircraft and in all Airbus fly-by-wire (FBW) aircraft, instead of a yoke in front of the pilot, there is a side-stick controller (SSC) outboard of the pilot seats with which to steer the aircraft; the motions of the SSC are analogous to the yokes. On aircraft with yokes, a pitch trim switch is installed under the thumb of the pilot to lessen control forces, making the airplane easily flyable throughout the flight envelope; Airbus FBW aircraft do not have a trim switch on the SSC because the flight control computers automatically trim the stabilizer throughout the flight after takeoff (this is referred to as "autotrim").

On the pedestal between the pilots are controls that both pilots need access to depending on who is doing the flying. These controls include the throttles (or "thrust levers" in some aircraft, though pilots invariably call them throttles), and controls for flaps and speedbrakes and/or spoilers. There are also master controls for controlling fuel to the engines that are activated during engine start and are again moved when the engines are shut down. Airliners have flaps on the trailing edge of the wing and a variety of leading edge devices (generally slats or leading edge flaps; there are technical differences between the two, but they are irrelevant to this discussion). At cruise speed, these devices are all retracted, but at takeoff and landing they are extended to differing degrees to increase lift when the aircraft is traveling at slower speeds. In most aircraft, important controls have distinctive shapes so that the pilots can identify them by touch alone: a prime example is the flap lever, which in most aircraft has a handle roughly the shape of a flap's cross-section. Generally, on the Captain's side (the left seat), there is a speedbrake / spoiler handle. Nomenclature varies a bit across manufacturers and phase of flight, but for the purposes of this discussion these are panels that pop up out of the top of the wing when activated. They function to destroy lift and increase drag. During flight, pilots use them to help descend more quickly when necessary, or to help slow down the aircraft. (Slowing down and descending at the same time is a very difficult task and requires a lot of practice and finesse.) On the ground, the spoilers deploy full up to help destroy all remaining lift, to slow the aircraft with additional drag, and to help deceleration by effectively putting more weight on the landing gear, enabling better braking. Some spoilers may only be used during landing (these are called, appropriately enough, "ground spoilers"). At the rear of the center pedestal are controls for radio tuning and switching, as well as interphone controls and public announcement controls. At the back of the pedestal are controls for rudder trim, which would be used to move the rudder to make control inputs easier in the event of an engine failure; most aircraft also have aileron trim controls there as well to slightly raise or lower one wing to achieve perfectly level flight (most Airbus aircraft don't have aileron trim, as it is all automatically controlled by the computers in the flight control system).

Immediately above the central pedestal there is an assortment of engine instruments and frequently lights or informational messages related to aircraft systems. Some aircraft have system synoptic displays and electronic checklists for both normal and abnormal operations located on screens in that general area. Also close by is the landing gear handle, with another piece of tactile feedback engineered in: at the end of the handle is a wheel shape (there is no mistaking it for anything else in the cockpit). Its function is highly logical: to raise the landing gear, the handle goes up, and the handle going down lowers the landing gear. In most aircraft (normally, though not universally, on the Captain's side), there are a set of standby flight instruments for the most rudimentary functions on the center panel. These basic flight instruments include aircraft attitude (pitch and roll relative to the horizon), airspeed, and altitude. These will be available on battery power, generally for 30–45 minutes, though systems are slightly different between aircraft types. These instruments provide only the most basic information to pilots, and if a crew were down to these only, it would be a dire emergency.

Immediately in front of each pilot are the primary flight instruments. Each pilot has an identical set of instruments, and each is powered from separate electrical sources and is fed from separate data sources for redundancy. Depending on the generation of the aircraft, the organization of the instruments may look a bit different. In older aircraft (e.g., DC-9 or B-747-200), mechanical instruments (e.g., gyroscopic attitude indicators and compass cards, etc.) and individual analog gauges (e.g., engine rotation, fuel flow, oil pressure, etc.) will be present. In more modern aircraft, some mechanical instruments will be mixed with cathode ray tube (CRT) displays for attitude and heading and map functions (e.g., B-757, B-767, many MD-80s), while in all new-build aircraft (e.g., A320, B-777) there are only CRT or liquid crystal displays (LCDs) combining everything into an integrated package (called a primary flight display and navigation display, or PFD/ND) in front of the pilots (see Figure 5.1). Regardless of the exact instrumentation present, the same basics will always be provided. The instrument panel places the attitude director indicator (ADI) at the center of the pilot's field of view. (In flying parlance, "attitude" refers to the aircraft's pitch and roll angles relative to the earth's horizon.) This instrument looks somewhat like a globe in that it has blue representing sky and a dark color (normally black or brown) representing ground. The horizon is the line between the colors. When the aircraft moves in the air, the actual pitch and roll of the aircraft are depicted by the aircraft symbol moving freely relative to the horizon line. This is the primary instrument pilots use to fly the aircraft in clouds. In older aircraft, the gyroscopes that provide data to the attitude indicator are mechanical; in newer models, this is accomplished by solid-state laser gyros without moving parts. On the ADI, there will be a set of command bars that can be turned on or off (they are normally on) that give the pilot cues about how to maneuver the aircraft in pitch and roll to fly a specified procedure (like an instrument approach). The autopilot(s) also get this information and can maneuver the aircraft during most phases of flight, in some cases even landing the aircraft. The controls for the autopilot and flight directors are generally on the glareshield above the instrument panel. On the right side of the ADI is the altimeter, which uses barometric pressure to tell the aircraft's height above sea level. There is also a radio altimeter that tells the aircraft's height above the ground, though in most aircraft this only functions below 2,500 feet above ground level. On the left side of the ADI, airspeed is displayed. Monitoring speed is critical for pilots, as flying too fast or slow can both result in disastrous mishaps. For every phase of flight, flap setting, and weight, there is an acceptable speed profile to be flown. Pilots are conscious not to exceed the acceptable speeds allowed for the operation being flown. Below the ADI there is a compass of some variety. In older aircraft, it was a simple horizontal situation indicator (HSI), which showed heading and displacement relative to a selected course. More modern aircraft replaced that with an electronic moving map display, and on the newest aircraft it gets its own display on the ND, which provides myriad route, course, track, and speed information in one concise display. Depending on the presentation, there will be a vertical speed indicator (VSI) as well to show pilots their rate of climb or descent (this is important, especially on instrument approaches). There will also be a clock located in a readily accessible position. Much depends on accurate timing from the execution of some approaches to engine start limitations.

In Figure 5.1, the screen on the left is the PFD, which shows the most important information for pilots, including: A, indicated airspeed; B, the ADI (formerly called the artificial horizon); and C, altimeter (note the VSI to the right of the altimeter). The right screen, D, is the ND, which displays heading, aircraft track across the ground, navigational aids, terrain, radar, and a moving map, among other things. E is the battery-powered integrated standby instruments (displaying attitude, altitude, airspeed, and heading; on some aircraft types, this is the only flight instrument to receive power when down to battery power). F is a radio magnetic indicator (RMI), which is a compass with needles that point at navigational stations. G is the multifunction control display unit (MCDU), the primary interface with the flight management computers and other features such as the Aircraft Communications Addressing and Reporting System (ACARS).

Figure 5.1. Captain's flight instruments on a newer Boeing 737-800. Photo: Leandro Luiz Pilch

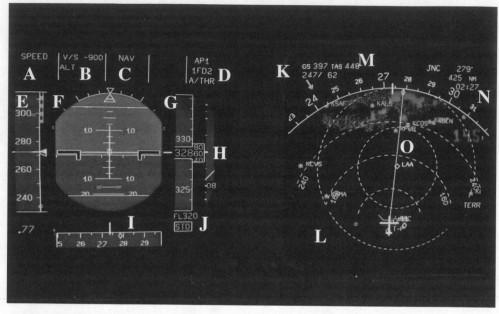

Figure 5.2. Primary flight display (PFD) and navigation display (ND) of an Airbus A320 during flight. Photo: Robert Hedges

The five top columns of the PFD (left screen, A–D) comprise the flight mode annunciator (FMA) and tells the pilots what the plane is doing at a glance; monitoring the FMA is crucial to flight safety (see Chapter 3). In this example, the aircraft is in a descent to flight level 320 (approximately 32,000 feet).

The first column (A) tells the pilots what thrust mode is active, in this case "SPEED," meaning the thrust is automatically being controlled to maintain the desired speed. The second column (B) shows the vertical mode, in this case "V/S," or vertical speed (see Chapter 8). In this case, the pilot has told the aircraft to descend at 900 feet per minute ("V/S-900"); note also that it says "ALT" below "V/S," because it is armed to automatically level off at a preselected altitude, in this case FL 320 (label J). The third column of the FMA shows the active lateral mode, in this case "NAV," meaning the aircraft will automatically follow the routing the pilots have programmed in the navigational system. Here the fourth column is blank, as Airbus uses that to display approach status information; since the airplane is at cruise, nothing will normally appear here. The fifth column (D) provides pilots with an overview of the status of major automation systems. In this case, autopilot 1 ("AP1") is engaged, both flight directors are on (one for each pilot "1FD2"), and the autothrust system is on ("A/THR"), which makes sense because column one revealed it is controlling the speed of the aircraft with thrust.

The tape on the left side of the PFD is the airspeed indicator. The present speed is read under the line in the middle of the display (in this case, about 272 knots. Note that the present Mach number is indicated numerically at the bottom of the tape (0.77 Mach). There is a small dot at about 235 knots. Airbus calls this "Green Dot," and it represents the best L/D_{MAX} speed for present conditions (it changes dramatically during flight owing to fuel burn). This is the best glide speed for the aircraft (see Chapter 7). Also note the single-column checkerboard pattern on the airspeed tape above 292 knots. It is red and black and provides a warning to pilots that they are exceeding a speed limit (exceeding this limit also causes an aural warning to sound); it changes according to several variables, including altitude and configuration of flaps and slats.

On the bottom of the primary flight display (PFD) is a compass showing heading and track across the ground to pilots. The heading is read under the center line (274°); this is also displayed under the compass rose at the top of the navigation display. The diamond to the right of the heading at (279°) shows the aircraft's track across the ground due to wind.

The right side of the PFD has two tapes; the larger one, G, is the altimeter. Here it shows the altitude as 32,860 feet; it also shows FL 320 below the tape, indicating that the plane will stop descending at FL 320 and hold that altitude. This makes sense because the flight mode annunciator (FMA) told the pilots that they were in a V/S descent. H is the vertical speed indicator and shows a descent of 800 feet per minute. Recall from the FMA that the pilot had requested a descent of 900 feet per minute, and here the reading for H agrees quite closely. These parameters will vary slightly depending on winds, turbulence, and the like, but the aircraft will correct back to the desired parameters as it adjusts during the flight. As mentioned before, J shows the altitude target; if it's below the current altitude, it will be below the altitude tape, if it's above it, it will be shown above the altitude tape.

"STD" indicates a standard altimeter setting of 29.92 inches of mercury (and equal to 14.696 psi), which airplanes use above a specified altitude (transition altitude; in the United States it is used passing through 18,000 feet, and altitudes at or above that height are referred to as "flight levels").

The navigation display (right) also has a lot of information available to pilots. As mentioned before, the compass heading is available under the top line in the center of the display (274°); the display can be configured as an arc (shown here) showing the pilots a forward view or as a complete circle with the plane at the center, which allows pilots to see what is behind them as well. In the arc mode, the aircraft symbol is at the bottom of the case, to the right of L. The map function is scalable and here is seen set on 320 nautical miles, which is typical at cruise. In the top left of the display, K, wind information, is displayed. In this case, the wind is coming from 247° at 62 knots. Note that the arrow graphically shows pilots the relative direction of the wind; at a glance, they can see it is a headwind and crosswind from the left. The information above the wind, M, is the current ground speed and true airspeed (TAS). The current TAS is 448 knots (which would be the speed across the ground in still air), but recall that the aircraft is experiencing a headwind, so the actual speed across the ground is 397 knots. The aircraft's route across the ground (O) is depicted as a green line (magenta in Boeing products) emanating from the aircraft symbol. Note here that it is not straight up but moves to the right as it gets toward the top of the screen, crossing the compass at 279°. That is the aircraft's actual track across the ground, and it makes perfect sense that it would be off to the right because the wind is coming from the left and pushing it. Radar and other data can be displayed on the ND, and in this case terrain is displayed, showing the highest terrain in various areas; airports have also been selected for display by the pilots in case an emergency diversion is required, they can see what is close by (in this case, KAMA, in Amarillo, Texas, appears to be the closest suitable airport; it's displayed just above L and is partially obscured by the ring legend displaying 160 nautical miles, or NM). In the top right corner of the display, information about the plane's next navigational waypoint is shown. In this case, the next point the plane is navigating to is JNC (Grand Junction, Colorado). It is on a track of 279°, is 425 NM in front of the aircraft, and the plane will get there at 02:27 universal coordinated time.

Cockpit Tour Continued

Outboard from both pilots will be an oxygen mask capable of providing 100% oxygen under pressure for the pilots to breathe in the event of a decompression or smoke in the aircraft. Smoke goggles are also present, sometimes combined with the oxygen mask, and sometimes separate. Outboard of the Captain is a small steering "tiller" that moves the nosewheel left or right for maneuvering the aircraft on the ground; some aircraft also have one outboard of the First Officer on the right side of the cockpit. Above the pilots there are many switches that largely control aircraft systems. Aircraft have complex systems that require much study and practice in the simulator to master. Although there are differences between aircraft, systems that are present in some manner on all aircraft are flight instrument control, electrical, hydraulic, fuel, fire protection and detection, pneumatics, air conditioning, and pressurization.

All airliners have either an older mechanical gyroscopic flight instrument system (which provides attitude and heading data to the pilot's instruments) or an inertial reference system, with laser gyros to provide the same information plus a lot of information that in combination with the aircraft flight management system (FMS) allows the plane to fly from any point on earth to any other point with no external navigation inputs.

Most airliners use large alternating current (AC) generators to power the high load electrical components on board (e.g., electrical hydraulic pumps, air conditioning cooling fans, etc.). Through a network of electrical "buses" that may have stepped-down voltages, lower-draw items are powered. Transformer-rectifiers convert AC into direct current (DC) power, which is used in much of the control and navigation circuitry. Additionally, aircraft batteries are able to power minimal electrical equipment for a specified time if all the generators quit. Some aircraft also have generators powered by the pneumatic system or by a ram air turbine, which is essentially a propeller that drops into the airstream in an emergency. These features are more common on aircraft that fly over oceans or are fly-by-wire, as electrical redundancy is more critical in those cases.

Airliners use hydraulic systems for actuation of most large components like flight controls, thrust reversers (some thrust reversers are pneumatically powered), landing gear, and brakes. Normally, each engine turns an engine driven hydraulic pump, and electrical pumps back those up. Some aircraft also have air turbine motors or air driven pumps that use high-pressure air to turn hydraulic pumps. Most planes also have reversible motor pumps of some variety, in which the pressure of one system can be used to turn a pump in another system. These systems can get extremely complex, but all boil down to a special fluid being pressurized to 3,000 psi (5,000 psi on a few aircraft like the A380) to move pieces against airflow or gravity that would be way too big to move in any other reasonable way. These systems are redundant because most modern aircraft require them to fly, while some older designs (e.g., DC-9, B-737) have a "manual reversion" mode allowing direct control via cables. Larger aircraft have flight control surfaces so large that pilots would be physically incapable of moving them even at low speeds. These aircraft have more hydraulic redundancy than older aircraft.

All airliners have a fuel system that ensures fuel is available to all engines. There are generally two pumps per fuel tank, and each is powered from a different electrical circuit. There are also "crossfeed" valves that can direct fuel from one tank to an engine it normally does not supply. This becomes especially important in the event of an engine shutdown during flight, as there is a maximum imbalance allowed between fuel tanks to ensure maneuverability and stability. Fuel temperature monitoring is provided and is especially important in long-range flights. Fuel can freeze in the tanks in some circumstances, and if fuel temperature gets too low, pilots will have to descend to warmer air (or possibly go faster, as the increased friction of the air passing over the wings can in some cases warm the fuel sufficiently to maintain a higher altitude). Fuel is used in some aircraft as a cooling agent for oil or other things, and in some aircraft, fuel heat can be selected to use high-temperature air to ensure there are no ice crystals that could occlude key fuel system components.

Airliners are equipped with fire detection and protection circuits associated with (at a minimum) the engines and cargo compartments. In most cases, two fire detection "loops" surround the structure being monitored. If one triggers, it will (in most cases) give a fault alert to the cockpit, but if both loops sense a fire, a fire warning will go off for that area. For an engine (or an auxiliary power unit), a handle or alternate control is provided, which when activated will cut off fuel to that engine, take its hydraulic pump offline, trip its electrical generator, shut off its bleed air, and arm the fire extinguisher bottles to be discharged. There are at least two fire bottles per engine available, and they are discharged in the cowling around the engine. Most engine fire warnings are due to a bleed air duct leak (this air is extremely hot and under high pressure). Modern airliners fly fine with an engine shut down. On three- and four-engine aircraft, Captains must demonstrate their ability to fly the aircraft safely with two engines shut down in the worst asymmetric thrust configuration (i.e., on a four-engine plane, two engines shut down on one wing; on a three-engine aircraft, the center engine and one other engine shut down). Unlike a single engine-out scenario, the two engine-out scenario is extremely challenging. Cargo compartments also have fire and smoke detection equipment and fire extinguishers installed, while some aircraft also have systems to monitor for wheel well fires or bleed air duct leaks throughout the aircraft.

Pneumatics, air conditioning, and pressurization all deal with the air systems on the aircraft. (Note that most of this discussion is not germane to the B-787, which has a novel air distribution system.) Air is tapped off the engines from ports called "bleeds" on the compressor stages of the engine. This air is extremely hot and under high pressure. It goes through a series of control valves and ultimately into air conditioning "packs" that use heat exchangers to lower the temperature and pressure of the air coming into the aircraft. The air coming out of the packs is extremely cold, and a small amount of the hot air is mixed back in to maintain the desired cabin temperature (newer aircraft are much more precise and user-friendly in this regard). The air flows in continuously and therefore flows out constantly, too, to keep the desired cabin pressure. There is a maximum cabin differential pressure for every aircraft, which normally allows the cabin to be pressurized to sea level up to a certain altitude, at which point the cabin pressure inside the plane also starts to rise. Most aircraft cruising in the 30,000- to 40,000-foot range will have a cabin pressurized somewhere between 6,000 and 8,500 feet. If the cabin rises above 10,000 feet, pilots get an alarm and must immediately put on their oxygen masks; if it rises above 14,000 feet, the oxygen masks in the cabin automatically deploy. (There are rare exceptions, as in modified aircraft that fly in and out of airports at extremely high altitudes. The best example is in airlines operating from La Paz, Bolivia, which must have modified systems because the field elevation is 13,325 feet above sea level, making it the highest international airport in the world.) Most newer aircraft recirculate a portion of cabin air, while older planes use a straight-through cycle. In most cases, the air in the cabin is exchanged approximately once every three minutes, which is much faster than in most buildings. In some aircraft, high-pressure pneumatic air also actuates certain hydraulic pumps or electrical generators; additionally, hot pneumatic air is used to deice or anti-ice engine air inlets and key components as well as portions of the wings' leading edges (and in some aircraft the horizontal stabilizer leading edges).

Simulators: Historical Perspective

Simulator training exploded during World War II. Attempts at accurate simulation did not occur until the military (always there first) linked the computer to the simulator. The first commercial flight simulator was built by Curtiss-Wright in 1948 for Pan American Airlines. Simulators continued to progress tracking developments of computers. As jets became larger and more complicated, a perverse situation developed where more planes were being destroyed practicing an emergency scenario than were destroyed by that scenario during normal flight operations—training flights were creating unsafe conditions.

For example, in 1959, a four-engine Boeing 707 crashed on a training flight, killing all five on board. During landing, the pilots reduced thrust on both left-side engines, simulating engine failure. Thrust to the remaining two engines was increased to maintain airspeed. The Captain trainee and instructor Captain lost control of the unbalanced plane at 1,000 feet, which was too low for recovery. The Federal Aviation Administration (FAA) later mandated practicing this maneuver at higher altitudes. Today, it's unthinkable to perform such a dangerous maneuver in actual flight instead of a simulator.

The 1970s saw an energy crisis, increasingly crowded skies, and simulation technology that was improving by leaps and bounds. In 1979 dollars, it costs $8,000 per hour to fly a Boeing 747. (This includes the fuel, 3,000 gallons per hour, but doesn't include the lost revenue of flying an empty plane.)

Full Flight Simulator Displaces All Training Flights

By the end of the 1970s, airlines were chomping at the bit for the FAA to allow experienced pilots to complete all training in a simulator, including upgrades (from First Officer to Captain), transitions (changing to a new plane), and recurrent training (required every six months to a year). Simulators had taken another leap in accuracy by incorporating more aerodynamic data, including ground effects (air flows over the wings differently close to the ground), landing gear effects, wind shear profiles, and so on.

United Airlines gathered data for 18 months with a Boeing 727. Recorded flight data and motion picture recordings of the flight instruments were compared to the simulation for a variety of circumstances. The data were duplicated in a second 727 operated by the FAA. In 1978, United was granted an FAA exemption for transition training evaluation.

New high-resolution cathode ray tubes could display full daylight for the first time with 32-bit processors in 1980 (Intel had just introduced its 16-bit processor for personal computers). Not coincidentally, the FAA established new rules for 100% training in simulators in that same year. Airlines had to establish procedures for FAA review and approval, and the FAA had to figure out how to certify the accuracy of the simulators. Finally, in 1982, United Airlines became the first airline to complete the process and began 100% training by simulator.

Required for the first time were full-daylight video (updated within 0.15 seconds after pilot inputs) and accurate noise and vibrations for takeoff, tire failure, turbulence, buffeting, rain, speedbrakes, and runway bumps. The full flight simulator also requires motion. The simulator was mounted on a platform attached to six actuators. The six actuators, in a process known as "acceleration onset cueing," provides acceleration in all six degrees of freedom: three linear accelerations and the three rotations of pitch, roll, and yaw. A row of three modern airliner simulators are shown in Figure 5.3.

Obviously, the platform cannot accelerate in any direction or rotation

Figure 5.3. Simulator in use at a major airline. Photo: Robert Hedges

without quickly reaching the actuator stops. Acceleration onset cueing provides the correct acceleration, albeit briefly, and then slowly decreases the acceleration. The platform is then gradually set to the neutral position, but at a rate below the threshold of human perception.

Initially, training instructors tended to be overzealous, and pilots began complaining of an unrealistic series of multiple emergencies followed in rapid succession—the pilots needed a chance to catch their breath. The process eventually developed a better rhythm. Today, simulator training curricula are highly researched and validated against real-world scenarios that have been challenging to other crews, and they are optimized to maximize useful training while minimizing unrealistic scenarios.

Simulator Accuracy

The simulator must match the plane during steady-state conditions of level flight, climb, descent, and constant turning. Test flights must also measure the plane's response to step changes of all the control surfaces (i.e., the rudder is suddenly changed 5° at a specific altitude, speed, and flap settings). Test pilots perform specific maneuvers such as turn to a new heading or climb to a new altitude in both the plane and the simulator. All the above is repeated with engine-out scenarios and other equipment failures over a wide range of airspeeds and altitudes. Simulators are so accurate today that they are used for accident investigation where pilot inputs and other flight control variables can be studied over a range of variables.

6

Turbulence

On January 10, 2008, an Airbus A319 took off from Victoria, British Columbia, at 6:25 a.m. bound for Toronto. At 6:45 a.m., air traffic control cleared Air Canada Flight 190 to climb from 35,000 to 37,000 feet. Three minutes later, the plane began a series of violent oscillations. Passengers on board described the experience in a variety of ways.

One passenger watched a friend fly "up to the ceiling and right down," and said the aircraft "went up and then sideways."[1] "All of a sudden [there were] three big drops . . . there was a crash. The cart tipped over and there was a lot of squealing. It was over and done with in 10 or 15 seconds."[2] The plane "went up sort of sideways and then came back down."[3] "You could see the wings going side to side, at one point, people thought it was going to go over, but it didn't."[4]

A 45-year-old nurse who helped treat the injured told her story. "I can't describe the screaming. No movie does it justice. There was a lot of screaming—a lot of crying. I thought it was over for me. I will admit I was saying my prayers because I really thought I was about to die. I said, 'Just take care of my family.'"[5]

The Flight Attendants had just started serving breakfast. One service trolley struck the ceiling, and another damaged the overhead bin. Minutes later, the Captain announced that the plane was being flown manually because of a computer malfunction. The Captain declared an emergency and diverted to Calgary. Other than being surrounded by 22 fire trucks and more than 10 ambulances—routine for an emergency landing (you don't want to be one fire truck short!), the plane landed uneventfully at 8:30 a.m. Nine passengers and crew were taken to the hospital, including six people who were evacuated, as a precaution, on stretchers with potential spine injuries. Eight were released later that day, the ninth the next morning.

Canadian investigators announced the next day that the plane experienced control problems, but it was too soon to determine whether the problems were mechanical, turbulence, or pilot error. The official investigation, released over two years later, reached a different conclusion.

At 6:33 a.m. (eight minutes after takeoff), air traffic control cleared the Airbus A319 to climb to 35,000 feet and informed the flight crew they were flying behind a Boeing 747. When the slightly faster Boeing 747 expanded its lead on the Airbus A319 to 8.1 nautical miles, air traffic control at 6:45 a.m. cleared the Air Canada plane to 37,000 feet, the same altitude as the 747. Just three minutes later, the A319, now 10.7 nautical miles behind the Boeing 747, began oscillating. The 151,100 lb (maximum takeoff weight) Airbus A319 had gotten too close to the 876,000 lb Boeing 747. In this case, too close turned out to be 10.7 nautical miles!

Wake Turbulence

Swirling tornadoes, an artifact of the physics of lift, occur at each wing tip, as shown in Figure 6.1. Airflow over the wings causes higher pressure below the wing and lower pressure above. At the wing tips, the higher-pressure air below the wing curls around and flows to the lower-pressure air above. The result is a pair of counter-rotating swirling air masses trailing behind the wing tips. Vortex wind speeds of nearly 200 mph have been recorded. This can cause a following plane to "roll" (or bank) its wings.

The wake turbulence moves down before leveling off. The vortices are never encountered above the generating plane and rarely more than 1,000 feet below. The vortices can persist up to 100 seconds in still air but will dissipate in 30 seconds with 6 to 12 mph winds, and faster still with higher wind speeds. Rare meteorological conditions allowed the remarkable image of the wake turbulence beneath and behind a Boeing 777 just after takeoff from London's Heathrow airport (see Figure 6.2).

Wake vortex strength is governed by the plane's weight, speed, and wingspan. Longer wingspans can better resist rolling caused by wakes. Separation rules were written based on plane weight and were recently changed to also consider wingspan. Smaller planes, with shorter wings, need more

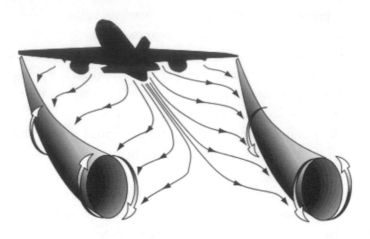

Figure 6.1. The wake turbulence behind and underneath all airplanes.
Federal Aviation Administration

Figure 6.2. The wake turbulence underneath and behind a Boeing 777 taking off from London's Heathrow Airport. Photo: Steve Morris

spacing behind larger planes. Far more complicated rules exist during take-off and landing. The rules have evolved to reflect the airplane's ability to resist a wake encounter. Plane weight categories are shown in Table 6.1; separation rules are given in Table 6.2.

The 151,100 lb Airbus A319 (111-foot wingspan) would be categorized "D." The 876,000 lb Boeing 747 (211-foot wingspan) is a "B." The stipulated separation for this combination of planes was 5 miles, more recently reduced to 3.5 miles.

Table 6.1. Aircraft Weight Categories

Category	Maximum Takeoff Weight (lbs)	Wingspan (feet)
A	>300,000	>245
B	>300,000	175–245
C	>300,000	125–175
D	<300,000	125–17
	>41,000	90–125
E	>41,000	65–90
F	<41,000	<125
	<15,000	all wingspans

Table 6.2. En Route Separation Rules

Leader	Follower Aircraft Separation (nautical miles) A	B	C	D	E	F
A	—	5	6	7	7	8
B		3	4	5	5	5
C				3.5	3.5	5
D						4
E						
F						

The strongest wake turbulence occurs when the plane is at its heaviest, at slow speed and clean wing configurations (i.e., flaps retracted), conditions that occur shortly after takeoff. During takeoff or landing, a following plane is restricted to wait two or three minutes (depending on circumstances) after the lead plane takes off or lands. Actual separation distance depends on the speeds and acceleration of the two planes involved. A large commercial jet rapidly accelerates to faster than 4 miles per minute during takeoff.

Previously called "prop wash," wake turbulence was not considered a flight hazard until the late 1960s. Serious research on wake turbulence began in 1969 with the introduction of the Boeing 747, a plane 2.5 times heavier than the next largest plane, at that time the Boeing 707. Back then, crude observations were made of planes flying past smoke-generating towers.

With many examples of fatal accidents, most wake turbulence mishaps (90%) involve small planes flying behind large commercial jets. There are even examples of small planes breaking up in flight behind a larger plane. For instance, in 2006, a six-seat Piper broke up flying behind a Boeing 737 during approach into Kansas City International Airport. A much heavier Boeing 747 could have a wake up to four times greater than a 737.

Planes flying close to the ground have less opportunity to recover from any extraneous and unexpected forces on the plane. This makes takeoffs and landings problematic, as demonstrated by the accident discussed next.

McDonnell Douglas DC-9 Crashes behind a DC-10

On May 30, 1972, a McDonnell Douglas DC-9, with a takeoff weight of 77,300 lbs, took off to qualify two Captain trainees. The pilots were practicing approaches and landings at Greater Southwest International Airport in Fort Worth, Texas. The pilot requested permission to land behind an inbound DC-10, with a takeoff weight of 330,000 lbs. The DC-10 was also a training flight performing touch-and-go landings. (Captain Hedges adds: This is a great example of why we now train in simulators: it's MUCH safer!)

At the time, when flying under instrument flight rules or visual flight rules with radar control, air traffic control (ATC) was required to provide a 5-nautical-mile or two-minute separation for any plane landing behind a "heavy" jet such as a DC-10. For planes following under visual flight rules without radar control, pilots were responsible for providing their own visual separation, the conditions applicable to the DC-9. In other words, if the following plane could see the plane in front generating the wake turbulence, there was no specified separation distance.

Just 39 seconds before the crash, the DC-9 was 55 seconds and 2.25 nautical miles behind the DC-10. 11 seconds before impact at an altitude of 60 feet, the DC-9 rolled gently counterclockwise as it passed over the runway threshold with the instructor commenting, "A little turbulence here."[6] The DC-9's right wingtip drifted into the clockwise wake turbulence of the DC-10's left wingtip, as shown in Figure 6.3A. This initial encounter with the wake vortex rolled the DC-9 counterclockwise. The DC-9 pilot applied countermeasures to level the plane. Unfortunately, when the DC-9 drifted farther into the wake vortex (see Figure 6.3B), the plane snapped over clockwise faster than it might have normally; the aileron correction made a moment earlier was still in place.

Unfortunately, this relatively benign first encounter with the vortex quickly led to a deeper and nonrecoverable penetration into the vortex core.

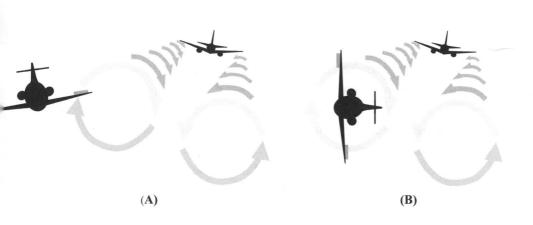

(A)   (B)

Figure 6.3. *A*, Initial encounter with wake vortex rolls the DC-9 counterclockwise. *B*, When the DC-9 enters the vortex core, the previous attempt to level the wings exacerbate clockwise roll.

Six seconds before impact, the instructor called for a go-around with take-off power. The right wingtip contacted the runway after the plane rolled approximately 90° to the right. The DC-9 continued to roll to an almost upside down position before striking the runway. The tail and fuselage separated upon impact and slid about 2,400 feet. The two pilot trainees, instructor pilot, and inspector from the Federal Aviation Administration (FAA) died in the impact and fire.

Before this accident, it was believed that wake turbulence was only a threat to small planes. After the accident, separation rules behind a "heavy" DC-10 were increased to 5 miles.

American Airlines Flight 587

On November 12, 2001, just one month after the tragedy of 9/11, American Airlines Flight 587 crashed in Queens, New York, less than 2 minutes after takeoff. The wide-body "heavy" A300, with a maximum takeoff weight of 353,500 lbs, took off 1 minute and 40 seconds behind a Boeing 747.

The Airbus was jostled by the 747's wake turbulence. This was not a control or structural problem for the large A300. The pilot, who had previously reportedly overreacted to wake turbulence, moved the rudder back and forth in an attempt to maintain the correct flight profile. Unbeknownst to the pilot, the rudder was structurally designed for full right or left rudder, but not for repeated and rapid back-and-forth rudder applications.

After full rudder deflection, a plane overshoots a little bit before it finds its new equilibrium heading. If the rudder is suddenly swung in the opposite direction, before the plane has found its equilibrium heading, the rudder will scoop out more air than it is designed to and will structurally overload the vertical stabilizer. The vertical stabilizer broke off of Flight 587 and the plane crashed, killing all 260 on board. Another way to explain the problem: the plane is not designed to fly sideways. The air blast of full rudder movement during excess yaw in the opposite direction exceeds structural limits.

The plane wasn't being flown the way it was designed; however, the pilots were not specifically instructed not to move the rudder back and forth as fast as possible, and in fact the pilots had received unrealistic and exaggerated training on the possible effects of wake turbulence on a heavy airliner in the airline's "Advanced Aircraft Maneuvering Program." As often happens after an accident, training procedures were changed. This accident is described in detail in *Beyond the Black Box*.

Airbus A380

Every time a significantly larger plane appears, there is increased concern about wake turbulence. And so it was for the Airbus A380, the largest passenger plane in the world, with a maximum takeoff weight of over 1.2 million pounds. Initial separation rules for the Airbus A380 for planes en route were 10 miles, double that of the Boeing 747. Airbus was not happy with the implication for future sales. The whole point of a larger airplane is to transport more passengers more efficiently at a lower cost. Increasing airplane spacing behind the A380 can reduce airport throughput and therefore the incentive to buy a larger plane. Airbus had tremendous incentive to justify reducing the separation rules. After a 3-year test program, including measuring over 1,000 wake encounters, Airbus justified reducing the A380 separation to that of the 747. There are no limits on an A380 flying behind any plane, including another A380, other than the normal requirements to minimize collision (the distance varies depending on the distance from airports).

Airbus recently performed 200 test flights with a smallish A318 (maximum takeoff weight of 150,000 lbs) flying behind a Boeing 747 and the heavier Airbus A380. Selected results are shown in Tables 6.3 and 6.4.

Airbus concluded that the probability of a severe wake turbulence encounter is remote because the following plane must enter a small swirl tube of about 20 feet in diameter and enter the wake tube at a 10° angle. "Even when trying . . . we did not manage to have a strong encounter every time."[7] Nevertheless, on some flights an unrestrained passenger will "bump on the ceiling"[8] (Airbus's words) at separation distances greater than the currently mandated 5 nautical miles. (Captain Hedges adds: This IS a problem! These variables are not well understood and rely extensively on the judgment of the crew. This is one reason pilots advise passengers to keep their seatbelts fastened anytime they are seated regardless of the status of the seatbelt sign.)

Table 6.3. Example Roll Response of A318 behind A380

Roll	Vertical Separation (feet)	Horizontal Separation (nautical miles)
46°	838	12.2
29°	1,015	18.1

Table 6.4. Example Roll Response of A318 behind 747

Roll	Vertical Separation (feet)	Horizontal Separation (nautical miles)
35°	432	5.32
31°	832	14.9
34°	1,038	15.8

Ground Effect

The so-called "ground effect" can be explained by a reduction of wake turbulence. If the plane is flown very close to the ground, the surface interferes with the wing's ability to stir up a large vortex. Drag on the plane is reduced, an effect usually only noticed during takeoff and landing.

The drag caused by wake turbulence is called induced drag. Although one should always be careful extrapolating results from one plane to another, *Aerodynamics for Naval Aviators* gives sample data for induced drag and ground effect. For a plane flying just above the stall speed, induced drag can be up to 75% of the total drag. Sample reductions of induced drag are 1.4%, 23%, and 47% for flight at 1 wingspan, ¼ wingspan and ¹/₁₀ wingspan above the surface, respectively.

In 1957, a four-engine C-97 (141-foot wingspan) was purposely flown just 100 feet above the ocean after a double-piston engine failure halfway to Hawaii. With an estimated five hours of remaining flight (and five hours and four minutes of fuel) the military plane with 67 passengers and crew ejected all luggage and freight, including mail bags. (With less weight the pilot can reduce the angle of attack, which also reduces drag.) With reduced power, the plane was flying only at 134 knots, about half the normal cruise speed of 260 knots. Flying closer to the surface would further reduce drag, but there are practical limits on how close the flight crew dares to approach the ocean. The C-97 safely landed in Hawaii.

Captain Hedges Explains: Wake Turbulence

The Air Canada A319, itself a large aircraft, proved to be no match for the wing-tip vortices of the much larger 747. Although separation criteria exist to keep smaller trailing aircraft safe from the vortices of preceding larger aircraft, the understanding of wake turbulence propagation is incomplete, leading to occasional surprise encounters with unseen vortices. In 99% of reported wake turbulence incidents, the encounters are abrupt and unexpected. Although over 90% of wake turbulence accidents occur to aircraft weighing less than 30,000 pounds, no aircraft is totally immune to the effects of wake turbulence, and the leading aircraft may not always be larger than the trailing aircraft. As a recent example, a Boeing 777 encountered the wake of another 777 at about 20 feet above the runway, requiring prompt pilot correction. As in many wake turbulence episodes, the wind was light (7 knots), preserving the preceding 777's vortex over the runway for a prolonged time.

Because the vortices are cones of rapidly rotating air behaving much like horizontal tornadoes, they are invisible to the eye unless there are clouds, fog, smoke, or some other atmospheric obscuration present, requiring the pilot in the following aircraft to mentally construct a picture of where they are in space for several minutes, all while occupied by the already busy task at hand (83% of wake turbulence accidents occur during takeoff or landing; most accidents occur during approach to landing below 200 feet above the ground). At any time a pilot may have to keep track of more than one set of vortices because of parallel or intersecting runways, and multiple arriving and departing aircraft. Pilots must be aware of weather conditions, particularly the airfield winds; the most adverse conditions are a mild crosswind and tailwind, which are both undesirable but together combine to keep a vortex over the runway for a long time (potentially over a minute, depending on the exact conditions). In general, pilots prefer to fly in light winds, but in dealing with wake turbulence, stronger and gustier winds are preferable as they tend to dissipate the vortices quickly.

Every aircraft creates two counter-rotating vortices, one from each wingtip, any time it is creating lift. These vortices settle toward the ground at about 300–500 feet per minute and in calm winds travel in opposite directions across the ground at about 5 knots. This is why a light crosswind (especially in combination with a tailwind) is the worst-case scenario for keeping a vortex over the runway for the longest period of time: in that scenario, the upwind vortex is held stationary over the runway because it has zero relative velocity to the ground. The net result is that the vortex simply sits over the runway until it dissipates. Therefore it is crucial that pilots be able to envision the locations of these vortices for several minutes.

The Air Canada A319 was fortunate to have encountered the 747's wake at 37,000 feet, rather than where most wake turbulence encounters occur, much closer to the ground with much less room to recover from the ensuing wild ride. Although wake turbulence encounters at altitude are not uncommon (though few are as severe as the A319 ran into), they are typically short lived.

Much more typical are encounters near the ground, like the DC-9 training flight in 1972 while following a much larger DC-10 around the traffic pattern at Greater Southwest International Airport. Although most wake turbulence accidents involve light aircraft, large jets are not immune to the hazards. In the final accident report, the National Transportation Safety Board (NTSB) found that "the hazard a DC-10 vortex imposes on a DC-9 is relatively as severe as the hazard which a B-727 or DC-9 vortex imposes upon a PA-28 or a Cessna 150 [small single-engine light aircraft]."[9] This accident is an ideal example for illustration of wake turbulence accidents in general. The weather and the surrounding terrain were the most adverse possible: a slight tailwind and light crosswind were present, which kept the vortex from the left wing of the DC-10 over the runway for a prolonged time, and the terrain around the airport was flat (higher or irregular terrain or structures helps break up vortex flow). The vortex stabilized about 60 feet above the runway on the runway centerline and was not visible to the crew of the DC-9. The crew was completing a challenging circling approach at the time of the accident and was only about 53 seconds behind the DC-10, substantially closer than would now be allowed (in 1972, there was substantially less understanding and regulation dealing with wake turbulence separation). The upset might have been prevented if the DC-9 had been farther behind the DC-10, but there was no guidance at the time to help pilots operating in good weather with visual flight rules (VFRs) determine what would constitute a safe distance.

Initially, as the DC-9 descended, it was above the vortices left by the DC-10, but eventually it converged with the track of the DC-10. There was a slight tailwind and crosswind from the left, which held the vortex generated by the left wing of the DC-10 over the runway. As the DC-9 approached the path of the DC-10, the initial encounter with the vortex was a modest roll to the left, which the DC-9 pilots countered with right aileron input, managing to keep the wings level. Although this reaction was reasonable in the view of investigators, it unwittingly set up the DC-9 crew for an "upset," in which the crew lost control of the aircraft. As the DC-9 descended through about 50 feet above the ground, the crew became more concerned about the wake turbulence, and the check airman ordered a go-around to gain altitude and get the aircraft out of the vortex. Unfortunately, at approximately the same time, the DC-9 penetrated the center of the vortex core and got into the other side of the vortex, which almost instantly initiated a severe roll to the right. With right aileron already displaced and a position just above the runway, the right wing rapidly descended and contacted the runway. Recovery was impossible, as "the velocity distribution of the vortex generated by the DC-10 airplane in the landing configuration induced a rolling moment on the DC-9 which exceeded the maximum lateral control capability of the DC-9."[10] This is a general truth in almost all wake turbulence accidents: for one reason or another, a (generally) smaller aircraft gets too close to a preceding larger aircraft, and the pilots of the trailing aircraft simply run out of aileron authority to counteract the rolling moment imparted by the vortex of the leading aircraft.

The wingtip vortices generated by aircraft are strongest when the aircraft are heavy, slow, and clean (i.e., no flaps or slats are deployed). Two of these criteria are met during takeoff and landing, with a heavy takeoff using minimal flap deployment generating the most severe conditions that most pilots encounter. Since a spate of accidents in the 1960s and 1970s, including the DC-9 in Fort Worth (the report on the DC-9 accident noted that between 1964 and 1971 alone, a wake turbulence encounter was a casual factor in approximately 120 aviation accidents), substantial research has been done into the nature of wake turbulence. A matrix that specifies how close different types of aircraft can get to each other on takeoff or approach has been widely adopted and has been largely successful in preventing accidents. Combined with more education and increased training emphasis, wake turbulence accidents for airliners are now rare. From a pilot's perspective, including wake turbulence in departure and approach briefings when following larger aircraft is now commonplace, and an early decision to go around when in doubt is also emphasized. Although in the case of the DC-9 accident the National Transportation Safety Board found that the most adverse set of conditions existed and that the lateral control authority was exceeded, they also noted that "the upset might have been averted had the pilot initiated a go-around at the first recognition of turbulence,"[11] a point that has since become a cornerstone concept in teaching pilots about dealing with wake turbulence encounters. Another key finding from this accident was the possibility of rapid roll reversal within a vortex; this is now a subject of more training than in 1972 (although the vortex encountered by the DC-9 was so severe that, possibly barring an early go-around, the aircraft would have been unrecoverable after the encounter commenced). This is an area that requires great experience and judgment on the part of aircrews, but passengers should be assured that far more is now known about wake turbulence, and current procedures are good at mitigating the risks of these encounters.

Smaller aircraft are most at risk from wake turbulence encounters, but even large aircraft are vulnerable to wake turbulence. American Airlines 587 is a unique case in that it encountered relatively modest wake turbulence behind a Boeing 747, and the Pilot Flying overcontrolled the aircraft with excessive rudder inputs, eventually detaching the entire vertical stabilizer from the aircraft and rendering it uncontrollable. It was not well understood by pilots outside the test pilot community at the time that full rudder deflection was safe, but alternating large rudder inputs were not accounted for in certification testing and could easily exceed the ultimate strength certification of the aircraft's structure. A surprise wake turbulence encounter can result in overcontrolling and pilot-induced oscillations (PIOs): the best course of action is to be wary of published wake turbulence separation criteria, consider the winds aloft, and stay away from (much) bigger planes. Wake turbulence is now an emphasis in airline training curricula, and it is considered an unusual accident that is fortunately unlikely to recur in airline operations.

Back to the wake turbulence encounter in 2008 for Air Canada Flight 190, mentioned at the beginning of this chapter. At 36,600 feet, the pilots reported three sharp jolts, similar to a car driving over speed bumps, followed by a roll to the right of nearly 6°. The autopilot tried to adjust and rolled the plane to the left almost 28°. The pilot disconnected the autopilot to fly the plane manually. Four rolls occurred with the maximum roll angle of 55° (flying upside down is a 180° roll). The event lasted 18 seconds, with maximum vertical accelerations of +1.57 G and –0.77 G; and maximum lateral accelerations of 0.49 (right) and 0.46 (left). The plane stabilized at about 35,500 feet, a drop of about 1,100 feet. The investigators concluded that the plane's response and effect on passengers were aggravated by pilot inputs, consistent with recent findings by Airbus. Airbus recommends that Airbus flight crews do nothing when they encounter wake turbulence at cruise altitude, as the autopilot should self-correct.

How far can a plane roll over? Surprisingly, a commercial plane can fly upside down,[12] and the wings will produce lift. If done correctly, the passengers will not even fall out of their seats—the centrifugal inertial loading will hold the passengers in place just like swinging a bucket of water overhead without spilling a drop.

G Loads

Normal gravity is 1 G. Standing on a scale in an elevator accelerating upward from 1 G or 32 ft/sec², the scale will read twice the normal weight, or 2 G. The scale reads normal weight plus the inertial effect from acceleration. Inertia is a property of matter that resists acceleration per Newton's Second Law, or $f = ma$. The inertial effect equals mass × acceleration. The inertial effect of accelerating up at 32 ft/sec² will result in a scale reading that shows the person weighs twice his actual weight. The person in the elevator accelerating upward is experiencing a force of +2 G.

Turbulence Is Costly

Turbulence is the most common cause of passenger injury—and the cause of much unnecessary passenger angst. From 1980 to 2008, US airlines had 234 turbulence accidents resulting in 298 serious injuries and three fatalities; 184 of the injured were Flight Attendants. At least two of the three fatalities were passengers who were not wearing seatbelts. About two-thirds of the accidents occurred above 30,000 feet.

It is estimated that turbulence costs airlines $100 million per year. Crew injuries can result in overtime and worker's compensation pay, and passenger injuries can result in lawsuits. Delays and cancellations may occur if injuries cause the plane to divert to a closer airport. (In 1999, one airline paid $2 million to 13 passengers for psychological distress. The passengers thought they were going to die after delays in turning on the "Fasten Seat Belt" sign.) Just the possibility of plane damage (albeit extremely rare) creates perhaps the biggest cost, as there is lost revenue for a plane taken out of service for inspection.

The Science of Turbulence

There are two types of flows: laminar and turbulent. Laminar flow is smooth and steady. Turbulent flow is random, erratic, and chaotic, with abrupt changes in velocity and direction. Nobel laureate Richard Feynman called turbulence the most important unsolved problem of classical physics.

Cigarette smoke illustrates the two types of flow and the transition from one to the other (see Figure 6.4). As the hot smoke rises, its speed increases. (The hot, less dense air is buoyant in the cooler surrounding air, just like a beach ball held underwater.) As the smoke's upward speed increases, the viscosity can no longer suppress turbulence. The cigarette smoke also illustrates a type of turbulence that affects airplanes known as convective turbulence. Provided the airplane is a speck of dust, it is easy to imagine the airplane being severely jostled by the cigarette smoke.

Another simple example is water flowing from a faucet. At low flow the water is quiet, glassy, transparent, and laminar. Injecting ink results in a straight-line streak of ink. At high flow the stream becomes chaotic, noisy, opaque, and turbulent. Even though the turbulent faucet flow is highly di-

Figure 6.4. Laminar cigarette smoke transitions to turbulent flow.

rectional (flowing downward in a torrent), ink injected will instantly mix, like the turbulent smoke. Not all turbulence is bad: for example, increased turbulence inside a piston engine enhances combustion through better mixing of the air and fuel.

The two types of flow were first studied scientifically by Osborne Reynolds in the nineteenth century. Airflow in the atmosphere is not exactly the water flow in a pipe studied by Osborne, but a description of his original experiment gives some insight into the on-again, off-again whimsical nature of turbulence, and why turbulence still defies precise scientific prediction.

Osborne injected dye into liquid flow inside a glass pipe. Initially, at low velocities the flow was laminar. The dye moved in a straight line along the length of the pipe. With increased speeds, small bursts of turbulent puffs formed and quickly dissipated. At still greater speeds, the turbulent puffs persisted for longer periods until eventually fully sustained turbulent flow occurred. Injected dye was then completely broken down into a churning turbulent mixture.

Turbulent flow is composed of eddies, also known as vortices. The vortices continually form and break down into smaller vortices. The eddies create the random velocity fluctuations capable of slamming around an airplane. The size of the vortices affects the plane's response. For large vortices, the aircraft can adjust pitch angle and airspeed to smoothly ride the waves. Smaller vortices excite rapid pitching and vertical motions of the plane. Even smaller vortices can excite a flexible response of the plane, that is, wing bounce or twist. (Passengers are often alarmed to see the wings tips bouncing a few feet. This movement should be compared to the structural certification tests; a Boeing 777 wingtip, for example, can flex 25 feet before structural failure.)

Even in a carefully controlled laboratory environment, the onset of turbulence is not a precisely defined repeatable process, and for that reason different textbooks will predict onset at different flow speeds. Assuming an experiment is designed with a statistical start of turbulence at 200 ft/sec, a carefully designed experiment can delay onset up to 2,500 ft/sec. For that experiment, it would be correct to say that turbulence begins somewhere between 200 and 2,500 ft/sec. It is now believed that laminar flow is stable for all flow rates. A disturbance is required to trigger turbulent flow. For pipe flow, the disturbance can result from surface roughness, random room vibrations, heat transfer, a crooked pipe, and even a loud noise. As the flow velocity increases, a vanishingly smaller disturbance is required.

A mechanical analogy is driving a car over a bumpy road. The car's suspension consists of springs and viscous dampers (i.e., the shock absorbers) that dissipate the car's bumpy oscillations. If there are no shock absorbers, the car's oscillations persist after the bumps have passed. A fluid's viscosity is like the shock absorbers, resisting the onset of turbulence. (It is difficult to make thick, viscous syrup become turbulent.) As the car's speed increases, its suspension system will not dissipate enough energy to adequately damp the oscillations, and the oscillations (i.e., turbulence) persist.

Incidentally, the distinction between laminar and turbulent flow has important medical applications. Almost all blood flow in the body is laminar. A heart murmur occurs when a heart valve does not properly open or close, creating noisy turbulence that can be heard with a stethoscope. Turbulent noise also helps measure blood pressure. A blood pressure cuff squeezes the arm and stops all blood flow. The pressure in the cuff is slowly released until noisy turbulent flow is heard through the restricted artery (the first pressure reading). As the pressure in the cuff is further released, the flow returns to quiet laminar flow (the second pressure reading).

Sources of Atmospheric Turbulence

Thunderstorms are dangerous sources of turbulence, but they are avoidable with modern onboard weather radar (no commercial jets have been destroyed by a thunderstorm in the modern jet era[13]). Wind shear, resulting from microburst downdrafts that flow outward, causing rapid variations in horizontal wind speed, is equally dangerous. Wind shear catastrophes have been mostly eliminated with the advent of Doppler radar.[14] (Captain Hedges adds: Also thanks to LOTS of pilot training, improved simulators, and onboard wind shear detection systems.) Remaining sources of atmospheric turbulence can be broadly categorized as mechanical or convective.

Mountain waves are a form of mechanical turbulence caused by wind blowing over mountains. Mountain waves are often compared to the turbulence in a river flowing over rocks in a section of rapids. If the wave amplitude is large enough, the waves become unstable and break, much like ocean waves in a surf, forming one type of clear air turbulence. The mountain forces the wind to move up. Surprisingly, some pilots have reported mountain waves as high as 60,000 feet, and hundreds of miles downwind from the mountain.

Clear air turbulence is also formed near large wind shifts, air mass collisions, or collisions in the jet stream. Unfortunately, clear air turbulence cannot be seen visually or with conventional radar.

Convective turbulence is as simple as hot air rises (or cold air sinks) and is similar to rising cigarette smoke. The most violent updrafts and downdrafts are in thunderstorms and are easily seen by radar because of water's high radar reflectivity. The problem is clear air (and unseen) convective turbulence. Convective turbulence can also be associated with fronts, which are collisions of air masses of colder and warmer air.

Clear air turbulence will also occur on the edges of the jet stream, where high-speed air mixes with surrounding low-speed air. The jet stream is generally concentrated between 30,000 and 35,000 feet, with winds up to 200 mph and occasionally approaching 300 mph. Commercial jets like to take advantage and fly with the jet stream as a tail wind when possible. Typically there is one jet stream across the United States, but there can be as many as three. Jet streams are thousands of miles long, hundreds of miles wide, and typically less than 3 miles high. Jet streams can also be discontinuous. The jet stream will move around seasonally and with weather systems. (Captain Hedges adds: We get a new jet stream forecast every six hours, as it can move around dynamically throughout the day.)

One of the worst cases of jet stream turbulence occurred on December 28, 1997, when United Airlines Flight 826, a Boeing 747-122 en route from Tokyo to Honolulu encountered severe clear air turbulence over the Pacific Ocean. At 31,000 feet and about 1 hour and 40 minutes into the flight, the plane experienced wave action, a type of turbulence where the eddies are so large the plane appears to "surf" up and down on a wave. The Captain considered this a possible precursor to turbulence and turned on the seatbelt sign. Announcements were made in Japanese and English. The pilot also radioed a plane ahead inquiring about the air, and the response was that it was smooth sailing.

Shortly thereafter, the plane experienced severe turbulence. The crew reported that an overspeed warning alarm sounded, and the number four hydraulic system low-pressure lights illuminated. The Captain considered diverting to Midway Island, the closest airport if the plane was damaged; however, the severely injured passengers and Flight Attendants favored a return to Tokyo. The crew later determined that a bumped switch most likely caused the low hydraulic pressure. The Captain assessed the airworthiness of the plane and the medical situation. Coincidentally, there were two medical doctors among the 374 passengers, but they did not speak English. After 20 minutes, it was decided to return to Tokyo. The Captain used his emergency authority to turn off course and to climb 500 feet. Air traffic controllers quickly gave clearance to land in Tokyo.

None of the passengers receiving serious injuries were wearing seat belts. A medical officer (without the benefit of an official autopsy) believed the single fatality was caused by a fractured spine. A nearby passenger thought the fatally injured passenger struck her head on the ceiling. Additionally, 15 passengers and 3 Flight Attendants had unspecified serious injuries.

There was extensive damage to the internal furnishings, but there was no internal or external structural damage to the plane. The black box recorded a positive vertical acceleration of 1.81 G, followed six seconds later by a −0.824 G.

Dodging Turbulence

The dispatcher assigns flight plans. Airplanes typically fly as high as allowed by Federal Aviation Administration rules to minimize drag and fuel burn. This maximum altitude depends on the plane's weight and will vary on every flight. In fact, it will vary significantly during the flight (recall the plane's weight can be more than 40% fuel). As the plane burns fuel, it naturally tends to gain altitude. But controllers must precisely know the plane's altitude. For that reason, planes fly at fixed altitudes, but may on long flights increase their altitude at discrete points to optimize fuel usage. The dispatcher also considers tail- and headwinds, and, mindful of delays, routes planes around storms and turbulent air. Planes rarely fly below 28,000 feet. Flying lower requires more fuel, which in turn requires still more fuel to carry around the additional weight of the additional fuel. The dispatcher typically inputs weather data, takeoff weight, and fuel data into a computer program to optimize all the variables.

Turbulence as Defined by the Federal Aviation Administration

Light turbulence barely registers. With moderate turbulence, unsecured objects move about and passengers feel strains against their seatbelts. Walking and serving food is difficult during moderate turbulence. Severe turbulence forces occupants violently against their seatbelts, and aircraft control is momentarily lost. During extreme turbulence, the aircraft is practically impossible to control. Extreme turbulence may cause structural damage. Historically, the level of turbulence has been assigned based on the pilot's recollection against the above criteria.

More recently, modern commercial jets have an accelerometer that measure G loads. The National Weather Service uses the following changes in vertical acceleration to quantify turbulence: light turbulence, less than 0.5 G; moderate turbulence, 0.5 to 1.0 G; severe turbulence, 1.0 to 2.0 G; extreme turbulence, greater than 2.0 G. For the same force, a big plane will be accelerated less than a small plane. Also, the acceleration sensor is near the plane's center of gravity (CG). Rotation about the CG can add to the plane's acceleration on the ends of the plane. In other words, individual passengers may experience more acceleration or G loads than recorded by the sensor.

Two G's of acceleration will not harm the human body, a topic explored in depth in *Beyond the Black Box*. The problem is being out of position and being slammed to the floor (or into the ceiling), catching a limb, or hitting a moving object.

Extreme Turbulence

The National Transportation Safety Board database uncovers just one example of extreme turbulence for an Airbus or Boeing plane in the last 25 years. The Boeing 757 was en route from Vancouver to Costa Rica on April 7, 1993. The flight crew declared an emergency following loss of control after encountering extreme turbulence at 42,000 feet over Texas. The Captain had just turned on the fasten seat belt light as the airplane approached forecasted convective activity. The First Officer disconnected the autopilot as both pilots pushed down the nose to prevent the plane from stalling. Both generators went offline and all power was lost to all engine, flight, and navigation instruments. The Captain took the controls and flew the plane with the emergency standby instruments and declared an emergency. The First Officer started the auxiliary power unit (an additional jet engine in the tail that can power some systems) and began emergency procedures to attempt to regain electrical power. The crew detected the odor of an electrical fire, and the "equipment overheat" message appeared on the instrument panel. About five minutes after losing both generators, they came back online. The plane landed in Houston. No plane damage occurred, and no passengers or crew were injured.

With a single exception, no plane in the modern jet era has been structurally destroyed by thunderstorms or turbulence—clear air or otherwise. In 1966, a Boeing 707 was destroyed after the vertical stabilizer broke off. The plane experienced severe mountain wave clear air turbulence when flying near Mt. Fuji in Japan. Structural damage to the plane will result if full rudder is applied with excess yaw, as described in Chapter 3. Likewise, the British Overseas Airways Corporation (BOAC, a forerunner of British Airways) 707 experienced a gust load from the side that broke off the vertical stabilizer; as it left the aircraft, the vertical stabilizer collided with the left horizontal stabilizer, which also separated from the aircraft. The uncontrollable aircraft subsequently entered a spin and crashed.

Airplane control and maneuverability, as well as structural design, have progressed considerably since the 707 was designed in the 1950s. Although sometimes pilots are still surprised by clear air turbulence, considerable progress has been made predicting turbulence.

Turbulence and Loss of Control

Historically, there have been many examples of plane crashes initiated by loss of control during turbulence. The most recent airline examples occurred in 1963 and 1964. In 1963, a Boeing 720 took off from Miami and crashed after encountering turbulence at 17,500 feet, killing all 43 passengers and crew. In 1964, a Douglas Aircraft DC-8 crashed shortly after takeoff after encountering turbulence between 5,000 and 7,000 feet, killing all 118 on board; another DC-8 crashed in that same year in almost identical circumstances. Significant rule changes resulted in design developments that improved stability, maneuvering, and control. Today, turbulence remains a tertiary effect during some accidents. Something triggers the event (in the case of Air France 447 described in Chapter 4, it was a frozen pitot tube), the plane gets tossed around in ordinary turbulence (a common occurrence), and the pilots for whatever reason get disoriented (a rare event) and lose control of the aircraft. Turbulence can even cause a stall, but the flight crew is expected to correctly respond; Captain Hedges reviews a recent example below.

Predicting Turbulence

Atmospheric turbulence can be predicted, but like any weather prediction must be taken with a grain of salt. (Captain Hedges adds: We take turbulence forecasts seriously, and they are generally good. We tend to err on the conservative side; when in doubt, we leave the seatbelt sign on.) Careful scientific study and computer models predict weather (and turbulence) trends and patterns with increasing accuracy. The models are stuffed with surface and satellite data. For upper atmosphere data, balloons are launched from 92 sites twice daily. (The balloon can drift for up to 2 hours and 125 miles.) Weather prediction took a quantum leap forward with NEXRAD, a network of 160 Doppler weather radars completed in 1992–1997. Doppler radar can see wind and rain inside clouds (not clear air turbulence) and adds to the data set supporting all weather predictions. Doppler radar can see about 140 miles and is not to be confused with terminal Doppler weather radar, a more sensitive radar (55-mile range) currently deployed at 45 major airports in the United States to forecast microbursts and associated wind shear.

Everything (hardware and software) slowly gets better, as improved weather (and turbulence) predictions remain an area of ongoing research. For example, beginning in 2008, Doppler radar resolution was increased and the range expanded to 250 nautical miles. New algorithms take the same data and process that data differently to make better predictions. Rockwell Collins and Honeywell, both major avionics suppliers of next-generation radar are competing head to head with onboard Doppler radar that claims enhanced detection of convective turbulence, not clear air turbulence. It remains to be seen how well Rockwell's MultiScan Threat Track and Honeywell's IntuVue perform. These new radars are expected to be deployed on the newest Boeing 737 MAX and 777X and Airbus A320neo.

Surprisingly, turbulent air persists for long periods of time. The most meaningful predictions of turbulence have always been real-time live reports from pilots. Pilots would manually report to dispatchers and air traffic control their assessment of severity, and dispatchers would disseminate the information. More recently, the process has become automated. Turbulence is measured with accelerometers on the plane and then automatically reported and disseminated. The automatic systems are trending up, with over 1,000 airliners in the United States so equipped today.

Captain Hedges Explains: Dealing with Turbulence

Nobody likes turbulence, but most turbulence is operationally merely a nuisance for pilots and has no safety implications for the aircraft. Moderate turbulence strikes no fear into pilots, as they will experience this level of turbulence for a few out of every 1,000 hours they fly. It usually lasts for no more than 10 or 15 minutes at a time, but occasionally may last for several hours and is often intermittent. This doesn't mean that pilots don't take turbulence seriously, as most injuries on airliners every year are due to turbulence. Pilots seek "ride reports" from company personnel, other aircraft, and air traffic control to ascertain whether another routing or altitude would yield a better ride, as this sort of turbulence will unsettle even some regular travelers. In moderate turbulence, an aircraft may change altitude typically around plus or minus 10 or 20 feet. No action is required by the pilot to control the aircraft, but the flight crew may decide to try a different altitude if the turbulence persists. In the cases of long-lasting turbulence or areas of mountain wave turbulence, sometimes a deviation from the planned route will be made. Mountain wave turbulence is caused by air passing over mountains; it can be quite serious and exist for thousands of feet in altitude and hundreds of miles downwind of the mountain range in question. If you are a frequent traveler, you may have noticed that the ride is frequently turbulent at airports close to (and especially downwind of) mountains, with Denver being an obvious example in the United States.

Severe turbulence is extremely rare. In a career of over 15,000 hours of flying time, I have experienced severe turbulence for about five minutes in total, most notably in a DC-9 circumnavigating thunderstorms embedded within clouds. While deviating around these storms, the aircraft had a radar failure that made us dependent on the much less accurate weather radar of the air traffic controller and reports from a preceding aircraft, which probably resulted in getting closer to a thunderstorm cell than desired. Severe turbulence is extremely uncomfortable but is not dangerous to the aircraft itself. Having said that, thunderstorms are full of other hazards besides turbulence that are potentially dangerous, which is why they are avoided. In severe turbulence, the aircraft may have altitude changes of possibly 100 feet, but nothing like the thousands of feet passengers frequently report to the media.

Even though severe turbulence is extremely unlikely to harm the aircraft (recall that the United Airlines 747 that had a fatal encounter with turbulence received no structural damage), severe turbulence is extremely hazardous if aircraft occupants are not wearing their seatbelts. Fatalities and severe injuries in turbulence almost always involve people who are not seated with their seatbelts fastened. In the case of the United 747 over the Pacific Ocean, a passenger who was not wearing a seatbelt was killed (and others were injured). Flight Attendants are vulnerable, as they can be out of their seats with the seatbelt light illuminated with the Captain's permission. A Boeing 767 clear of weather encountered 40 seconds of severe turbulence over North Carolina on July 11, 2005, in the middle of an otherwise smooth ride. The turbulence was so sudden and severe that the pilots were momentarily unable to grab the controls; when they could, they turned on the seatbelt sign and inquired about the safety of the passengers and crew. The plane ultimately diverted to Raleigh with an injured Flight Attendant. The aircraft had earlier passed through a stormy area, but left the weather far behind; with a smooth ride at 37,000 feet and reports from air traffic control (ATC) to expect more of the same, the crew turned off the fasten seatbelt sign only to subsequently encounter severe turbulence. This type of turbulence, clear air turbulence, is particularly insidious. Current radars are unable to detect clear air turbulence (the crew stated they were monitoring the radar and no weather was present on it) because they are dependent on the high reflectivity of water droplets in the air; if there is no water in the air, they cannot detect turbulence (even for contemporary radars with a "turbulence" mode). This is why it is so important for passengers to keep their seatbelts fastened whenever they are seated, even if the ride seems smooth. This example is more severe than most, but nothing in turbulence prediction is foolproof.

There is a category of turbulence beyond severe: extreme turbulence. In over 15,000 hours of flying I have never experienced it, nor do I want to. I do not personally know anyone who has experienced it. Extreme turbulence is just that; it can be bad enough to structurally damage aircraft. Fortunately, the most likely place to encounter this type of turbulence is near or in a thunderstorm. Because thunderstorms are huge and conspicuous on radar and to the naked eye, the probability of encountering extreme turbulence is remote, as all airlines have policies to avoid thunderstorms. If ATC can't provide a deviation around a thunderstorm, Captains are authorized to use their emergency authority to go around the storm if necessary. This is not common, as ATC is normally aware of where the weather is and plans for pilots needing to go around it.

Although turbulence is generally not particularly hazardous to aircraft, it can serve as an initiator of an accident sequence or a link in the chain resulting in an accident.[15] After losing airspeed information, an additional distractor for the crew of Air France 447 (see Chapter 4) was the turbulence experienced during the episode; likewise, the 2009 incident involving a Europe Airpost 737 arriving at Antalya, Turkey, involves turbulence that complicated an arrival to the point that the aircraft stalled. Unlike AF 447, however, the 737 crew was able to diagnose the stall and recover the aircraft safely.

The Europe Airpost crew was using the radar and was deviating around a thunderstorm, both of which were in accordance with good practices. ATC requested the crew to slow the aircraft, and they selected 220 knots, 10 knots above the minimum maneuvering speed for their weight and clean configuration (flaps up). They were operating in an area of known mountain wave and turbulent conditions, so a good choice would have been to either maintain a higher speed or begin to extend flaps to reduce the required minimum maneuvering speed: a 10-knot margin in turbulence is slim. During the arrival the turbulence increased rapidly, the autothrottles initially retarded, and the aircraft began to slow; a crewmember overrode the autothrottles, but the engines did not accelerate symmetrically. In quick order, the aircraft got slow (as slow as 181 knots), began to roll to the right rapidly (reaching 102° right roll; the wings were well beyond the vertical), reached the critical angle of attack, and stalled. A rate of descent in excess of 12,000 feet per minute developed during the upset, and the crew struggled to recover. This was made more disorienting by virtue of being in clouds. Although like Flight 447 the crew initially applied nose-up elevator control in reaction to the high rate of descent, they quickly realized that the aircraft was stalled, added power, rolled the plane back to the left, and lowered the nose until the stall was broken and the wing was flying once again. At that point they recovered from the dive, composed themselves, and safely landed the aircraft. This was an extremely challenging situation, and thankfully the aircraft and its occupants were unhurt. Because of this and other upsets, pilots now receive regular practice with unusual attitudes in the simulator. In all cases where the aircraft is stalled, it is paramount to first unstall the wing by using nose-down elevator force; adding power and rolling wings level will then permit the pilots to regain control of the aircraft, provided that the plane was high enough above the ground before the event began, as altitude can be lost at very high rates, as was the case in the Air France Flight 447 accident.

7

The 168-Ton Glider

The Airbus A330 left Toronto on August 24, 2001, with 306 passenger and crew bound for Portugal. The 441,780 lb plane took off at 8:10 p.m. with 104,500 lbs, or about 12,600 gallons, of fuel. This included a 5% reserve required per regulations plus an additional 16% added for potential inflight rerouting and tankering.[1] Unbeknown to the flight crew, a serious fuel leak began almost four hours into the flight. Surprisingly, the plane's electronic centralized aircraft monitoring (ECAM) system did not, at that time, directly monitor potential leaks. Instead, standard procedures called for the crew to periodically check the fuel on board for discrepancies. The fuel check was done six times on this flight, with the last occurring about 20 minutes after the leak started. No fuel anomalies were recorded by the crew. The leak started slowly before growing rapidly.

The A330 normally uses about 1,300 gallons of fuel per hour during cruise. The fuel leak at its peak leaked about 3,600 gallons per hour, a rate of a gallon per second. The first indication of anything amiss was, of all things, engine oil temperature anomalies occurring about 25 minutes after the fuel leak began. The engine oil temperatures were indirectly related to the fuel leak but, of more significance to the story, did not make sense to the pilots.

Normally the crew has a written procedure for every contingency imaginable. If the existing procedures don't exactly fit the situation, plan B is to call the engineers at the airline's headquarters for advice. In this case, the engineers were equally baffled. At the time, nobody considered a fuel leak. The flight crew was thinking "computer error," as instrumentation errors can provide erroneous and unexpected cautions and warnings that otherwise do not make sense. (Similarly, in 1970, mission controllers at the National Aeronautics and Space Administration in Houston initially thought the malfunctions experienced by Apollo 13 were due to instrumentation error; they were actually the result of an explosion in an oxygen tank.)

When the magnitude of the leak was finally determined, the amount of missing fuel appeared far-fetched and unbelievable. The crew's first response was to check the flight's documentation for a harmless clerical error. In fact, until the first engine flamed out, the flight crew continued to believe the missing fuel was an illusion and caused by computer error.

We pick up the blow-by-blow nearly five hours into the flight, at 5:33 a.m. Coordinated Universal Time. At that point the computer displayed a 6,600 lb fuel imbalance; there was 6,600 lbs less fuel in the right wing tank, a staggering difference. The most common cause for a fuel imbalance is not a leak, but rather the engines burning different amounts of fuel or one engine shutting down and not burning any fuel. The computer presented the fuel imbalance as an "advisory" white message as opposed to a more serious amber or red message. (Amber indicates loss of system redundancy or system degradation with no immediate effect on flight safety, and red indicates an urgent issue that requires prompt crew attention.)

Today, pilots use the autopilot to fly the plane most of the time, especially on long flights over oceans. If the pilot had been flying the plane manually, the fuel imbalance could gradually roll the plane, and the problem would be easily diagnosed. Instead, the autopilot adjusted the controls to keep the plane straight and level, masking the problem.

A serious fuel imbalance almost never presents itself in actual flight, but this scenario is frequently practiced in the simulator during "engine-out" procedures. At 5:36, the crew began to balance the fuel in the wing tanks, following a procedure from memory. They transferred fuel from the left side to the leaking right engine, which made things worse. If they had read the procedure for a fuel leak, they would have learned of a warning not to transfer fuel when a leak is suspected. Shortly thereafter, the crew discovered that there was more than 2,200 gallons of fuel missing. After failing to find a clerical error related to fueling the plane, they began a diversion to a Portuguese military airfield in the Azores about 990 miles west of Lisbon at 5:45. Five minutes later, the Captain asked the head Flight Attendant to check the wings for a fuel leak. They checked and saw nothing.

Around 6:01, the head Flight Attendant was instructed to prepare the passengers for a water landing in about 40 minutes. The Flight Attendants conferred and began donning their life preservers. Beginning at around 6:13, as the right engine (number 2) flamed out, the passengers were instructed in English and French per Canadian law (recall that the flight took off from Toronto) for a water landing. The instructions continued for about 13 minutes as the passengers put on their life preservers.

The plane was safe to fly with one engine. In fact, regulations permitted this plane to fly for 150 minutes on one engine at 491 mph in still air or for 1,227 miles with no wind.[2] After running out of fuel, the second engine shut down at 6:26; the cabin lights began flickering and the public address system became intermittent. Cabin pressurization was gradually lost after the engines shut down, and the oxygen masks dropped about 5 minutes later. (Emergency breathing oxygen for passengers is generated with a powerless chemical reaction in the A330.) The emergency power quickly restored emergency cabin lights. The passengers, anticipating a water landing, began what was probably the most anxiety-provoking ordeal of their lives as they prepared for ditching. The plane continued its silent trajectory. How far would it glide?

The Touchdown

Much to the passengers' relief, the First Officer announced about 5-7 minutes before touchdown a "landing" as opposed to a "ditching" at sea. Immediately before the landing, the First Officer ordered "Brace, Brace, Brace" to the Flight Attendants, who shouted the command to the passengers, who assumed the brace position as the Flight Attendants had instructed them during their cabin preparations.

After a 19-minute unpowered 75-mile glide to Lajes Air Base in the Azores (a common alternate airport for flights from North America to Europe), the plane touched down hard at 230 mph (about 35 mph faster than recommended for a powerless landing) and 1,030 feet down the 10,000-foot long runway. The plane bounced back into the air and touched down a second time about 1,800 feet farther down the runway.

With limited power, the antiskid braking system was not available. The brakes locked and quickly abraded the tires; 8 of 10 tires blew out within a few hundred feet of the second touchdown. Small fires started in the left landing gear and were quickly extinguished by responding fire crews. (Fire trucks meet the plane on the runway whenever the pilot declares an emergency.)

The plane stopped with about 2,400 feet of runway to spare. The passengers began clapping and cheering. The Flight Attendants, anticipating immediate evacuation instructions, yelled for order; the passengers readily complied. Almost immediately, the Captain ordered an emergency evacuation down the slides. Some pilots, while hesitant to second-guess the flight crew after the fact, might not have elected to evacuate the plane. Fire hazards are reduced with empty fuel tanks, and passengers are frequently injured during an emergency evacuation.

Why Did the Plane Run Out of Fuel?

During a routine inspection on August 15, metal chips of an unknown origin were found in engine number 2's oil system's chip detector. Chips were found again on Friday, August 17, just six days before the incident. Even though the engine was operating smoothly, it was decided late that afternoon to preemptively replace the engine. Sixty components were required to be correctly changed out to install the replacement engine, including the starter, front and rear hydraulic pumps, connecting tubes, and other parts. Work began that night at midnight and was expected to be completed by noon on Sunday. The work was performed by a crew of four to six technicians working eight-hour shifts around the clock.

The replacement engine came without hydraulic pumps. On Sunday morning, the hydraulic pump casing removed from the existing engine would not fit on the replacement engine—it interfered with the fuel pump inlet tube. The technicians suspected changes in design layout between the replacement engine (which had been in storage for over a year) and the engine being removed from the aircraft. When reviewing the interference problem on the computerized *Airbus A330 Illustrated Parts Catalogue*, a service bulletin describing the problem (and the piping changes required to fix it) was referenced. The technicians could not directly access the service bulletin because of computer network problems. The service bulletin stated that two hydraulic and three fuel lines had been redesigned to solve the interference problem. The replacement engine had the old design hydraulic and fuel lines attached; the existing engine had the new design.

Rolls-Royce, the engine manufacturer, made the changes because excessive hydraulic pump vibrations had resulted in several leaks on this engine type. To eliminate the problem, Rolls-Royce redesigned the hydraulic pump casing. The new wider pump casing interfered with attached hydraulic and fuel lines, requiring their redesign.

Discussions between the lead technician and the airline's engine controller[3] about the interference problem concluded that the fuel lines attached to the replacement engine (the old design) must be replaced with the new-configuration fuel lines attached to the existing engine. The service bulletin, never accessed by the technicians, stated that the fuel lines and hydraulic lines must be replaced together as a set.

The interference problem with the hydraulic pump was solved when the fuel lines were replaced. But the hydraulic line then became difficult to attach. The technicians had incorrectly mixed the "new" fuel lines with the "older" hydraulic lines. The problem hydraulic line was an 11-foot-long flexible hose with rigid tubes on each end. The flexible hose, bent 90°, was used to isolate pump pulsations. The technician recalled having difficulty installing this line and had to hold the flexible hose in position while tightening the fastening nut on one end.

The work was finally completed on Sunday, August 19, at 5:30 p.m. The engine was inspected by the lead technician and again by a second inspector not involved with the work. The engine was started, run, and tested before being released from the shop.

Test installations after the accident showed that a 0.025-inch clearance between the hydraulic and fuel tubes was achieved by manually bowing the hydraulic flex line away from the fuel tube. But the hydraulic line expanded radially when pressurized to its operating pressure of 3,000 psi and contacted the fuel line. Seventeen flights later (67.5 hours of flight time), the rubbing lines caused a leak.

In another missed opportunity, a Rolls-Royce representative visited the hanger on Saturday, August 18. The Rolls-Royce representative was not aware of the technician's difficulty in accessing the correct service bulletin or any other problems. His offer to revisit on Sunday was not taken up. The

Canadian authorities eventually fined Air Transat $250,000 for not follow-ing written repair procedures and immediately suspended Air Transat's Extended Operations (ETOPS) authority until completion of a safety audit.

Engine Oil

Oil lubricates and cools the jet engine's main shaft bearings and the bear-ings in auxiliary equipment (e.g., fuel and oil pumps). Because the pressur-ized system leaks a small amount of oil past the seals, a jet engine typically consumes about 0.2 to 0.8 quarts of oil per hour. During normal operations, frictional rubbing within the oil causes heat generation. Excess heat will degrade the oil and must be removed in a heat exchanger. The oil param-eters for the two engines recorded about four hours into the flight are as shown in Table 7.1.

An incipient bearing failure will raise the oil temperature; an oil leak or oil pump failure will cause a pressure decrease. For all these reasons, the oil temperature and pressure are routinely monitored. The crew was confused by an apparent small leak that caused the pressure to increase and tem-perature to decrease, which was seemingly contradictory information. The crew called Air Transat operations for help; the maintenance personnel on the ground were confused too.

The fuel leak caused the flow rate of fuel through the heat exchanger to triple (Figure 7.1). Tripling the flow rate caused the heat exchanger to re-move roughly three times more heat from the oil. The oil, which normally exited the heat exchanger at 230°F, was chilled to 149°F. The cooler oil be-came thicker and about four times more viscous. The thicker oil was more

Table 7.1. Oil Parameters for the Two Engines

	Start of Flight	Four Hours into Flight	
		Left Engine No. 1	Leaking Right Engine No. 2
Oil temperature (°F)		230	149
Oil pressure (psi)		80	150
Oil quantity (quarts)	16.8	16.5	13.2

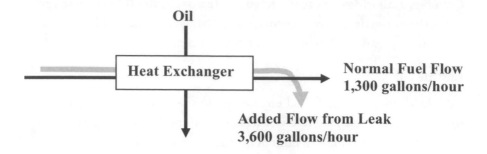

Figure 7.1. The lubricating oil is cooled by the fuel in a heat exchanger.

difficult to pump, resulting in a pressure increase to 150 psi. The increased pumping effort reduced the flow rate back into the oil reservoir; oil accumulated elsewhere in the system, and the sensor in the reservoir recorded less oil quantity in the reservoir.

The effects of various leak rates on the oil pressure and temperature are predictable by normal engineering calculations. Obviously, this scenario had not been thought through ahead of time. The oil parameters of the leaky right engine (number 2) were unusual, but not out of the allowable ranges. For that reason, the computer did not present a Warning or Caution but an advisory suggesting additional monitoring.

Why Didn't the Pilots Follow the Leak Procedures?

To land with fuel, the investigators concluded that the leaking engine needed to be isolated and shut off before 5:54 a.m. A variety of scenarios were considered in which the crew followed the various fuel leak procedures (Leak from Engine, Leak Not From Engine, Leak Not Located) beginning at 5:45. In all cases, the investigators concluded that the plane could have landed in the Azores with nearly 1,000 gallons of fuel had the leak diagnosis been made in a timely way.

The Captain later stated that he chose to disregard the applicable "Leak Not From Engine" procedure, which called for shutting off one engine (to isolate the leak) and descending to 20,000 feet. Anticipating a powerless landing, the Captain preferred to lengthen the glide by remaining at 39,000 feet (there is less drag and higher fuel efficiency at higher altitudes).

Why did the Fuel Leak procedure require reducing altitude to 20,000 feet? With one engine shut off, the plane does not have enough thrust to overcome the drag at the normal cruise speed. At a slower speed, there is less lift. The plane must find a new equilibrium between the basic forces of lift, drag, and thrust at a lower speed. This occurs at a lower altitude, where the air is denser. Compared to 39,000 feet, air at 20,000 feet is twice as dense.

Emergency Electrical Power

The A330 has three electrical generators driven by the jet engines and auxiliary power unit. There also is a hydraulically driven emergency generator as a backup. The A330 has a ram air turbine (RAT), a small wind-powered turbine that falls out of the fuselage to generate emergency electrical power. In case of emergency, it powers basic flight instruments and provides hydraulics for flight control surfaces. When the slats extend from the wings for landing, air flow to the RAT is partially blocked. At that time, a set of batteries supplying about 30 minutes of emergency electrical power are activated. On either RAT or battery power, only safety-critical electrical elements are powered.

Glide Path

The forces on a plane flying in straight and level flight are as shown in Figure 7.2. The lift must equal the plane's weight, and the thrust offsets the air drag. Because there are no net forces on the plane, Newton's law predicts zero acceleration, which means the plane is flying at constant speed.

The maximum lift-to-drag ratio for an Airbus A330 is 17.5. This value can be used to approximate the lift and drag forces on the plane when cruising. To economize fuel, the plane is designed to cruise at a specific speed that minimizes drag and maximizes fuel economy at any given altitude. An A330 at 36,000 feet cruises at approximately 540 mph. Assuming our plane took off weighing 441,780 lbs and burned about 20,000 lbs getting up to cruise altitude, the lift at 36,000 feet is approximately 441,780 lbs –20,000 lbs = 421,780 lbs. If the lift is 17.5 times greater than the drag, we estimate

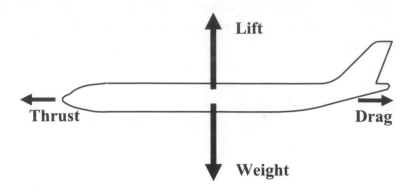

Figure 7.2. The forces on a plane flying in straight and level flight at constant speed.

the drag at 36,000 feet when cruising at 540 mph to be 421,780 lbs ÷ 17.5 = 24,100 lbs. To overcome this drag, each of the two engines must provide 24,100 lbs ÷ 2 = 12,050 lbs of thrust. This estimated cruise thrust agrees closely to the published cruise thrust for these engines (11,500 lbs).

If the engines flame out, there is no thrust—the plane is gliding. If the sum of the three forces acting on the plane (weight, lift, and drag) is in equilibrium, the plane is flying at constant speed, albeit in a descent (see Figure 7.3).

The combined effect of the lift and drag (i.e., the vector sum of the two) is shown in Figure 7.4. The combined lift and drag must be exactly equal and opposite to the plane's weight. If the lift force is 17.5 times larger than the drag force, then the angle between the combined lift and drag force and lift force will be 3.57°. From simple geometry, the angle between the between the glide path and horizontal line must also be 3.57°.

Because of similar angles, the triangle defined by the lift and drag forces is "similar" to the triangle defined by the horizontal and vertical glide paths. Similar triangles have the same ratio of any two sides. This means that for our Airbus A330,

$$\frac{\text{lift}}{\text{drag}} = \frac{\text{horizontal glide path}}{\text{vertical glide path}} = 17.5$$

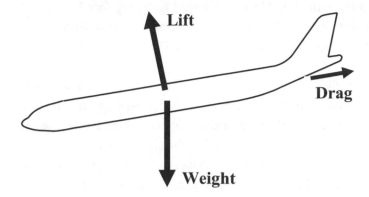

Figure 7.3. The plane glides at constant speed if the sum of the three forces (lift, weight, and drag) equate to zero.

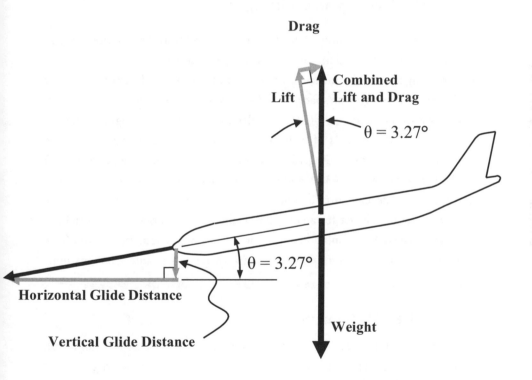

Figure 7.4. For lift ÷ drag = 17.5, the force balance triangle and glide ratio triangle are similar, with both having an angle of 3.27°. A larger angle is drawn here for clarity.

The Airbus A330 will glide 17.5 feet forward for every foot it drops, if flown optimally. Both engines flamed out at an altitude of about 34,500 feet. Theoretically, the plane can glide 34,500 feet × 17.5 = 603,750 feet, or 114 miles.

Air Transat 236 glided to a landing at about 230 mph. This compares to an estimated stall speed of about 140 mph with the flaps up and 116 mph with the flaps down for this relatively light plane with zero fuel. Once the pilot lowered the landing gear and extended the slats, the drag increased dramatically, and the plane's ability to glide was severely compromised. The lift-to-drag ratio for a similarly sized McDonnell Douglas DC-10 with the flaps and landing gear down was about 6.6.

Captain Hedges Explains: Gliding

It may seem counterintuitive, but airliners spend a lot of time gliding. In fact, the ideal, most fuel-efficient descent is to fly to a precomputed point and retard the engines to idle, descending to a point on the final approach. In the real world of congested airspace, this ideal rarely happens, but parts of almost every descent are made at idle thrust. Although even in idle engines produce some residual thrust, it is of little real consequence, and for all intents and purposes the aircraft is gliding. All airliners, from the smallest regional aircraft to the behemoth Airbus A380, glide. Gliding is a useful tool as a planned part of an aircraft's descent profile, and it also occurs when an aircraft loses the ability to produce thrust for any reason. Although it is extremely rare to have all engines cease functioning in an airliner, it has happened, most often from fuel exhaustion or fuel contamination, but also due to volcanic ash ingestion.

How Much Fuel Is Required?

It's complicated. Part of a pilot's job is fuel planning. This task is done well before the flight, and in airline operations, this job is shared with a licensed flight dispatcher who has knowledge of the route, weather, winds, and so on. The actual amount of fuel required can be complex to calculate because it depends on a variety of factors. It varies from country to country (different countries and regulatory regimes have different standards and nomenclature), and airlines impose their own rules. For a US-registered aircraft to cross an ocean, there needs to be sufficient fuel to reach the intended destination, fly an approach, fly a missed approach (go-around), fly to an alternate destination, fly an approach there, land, add a value equal to 10% of the en route time (including approach and landing), plus the amount needed to hold at the alternate for 30 minutes, fuel to taxi in and out from the gate, plus any fuel to cover known or expected contingencies (long taxi-out delays, thunderstorm deviations, known arrival delays, holding, etc.). Other details can alter this number, but this is a general starting point. In fact, the rules are so rigid on fuel requirements that occasionally a paying passenger will be bumped off a flight in favor of extra fuel. (This could occur if a greater than normal headwind is forecast, for instance.)

It is always prudent for a Captain to review the dispatcher's initial fuel planning, and sometimes to request that more fuel be loaded based on his or her experience with the specific route and weather; requesting extra fuel is common in the summer—planes often operate in areas with forecast thunderstorms, for instance—because increased fuel for course deviations or holding patterns gives more time and options other than a rapid diversion to another airport. This request isn't unlimited, as the aircraft can't take off with so much fuel that the flight will plan to land above the maximum landing weight.[4] Often fuel will be "tankered" from somewhere with lower fuel costs to another country (or city) where the price of fuel is high. Although it takes more fuel to carry the tankered fuel, tankering can save thousands of dollars per trip (dispatchers use sophisticated software to figure out the costs).

Lessons in Fuel Management

Safety investigators speak in terms of a chain of events, or an error chain, when investigating accidents: it's rare that a single event causes an accident (or incident), and Air Transat 236 was certainly no exception. The A330 is an automated aircraft with computer displays showing a lot of information—including systems diagrams, checklists, pressures, quantities, and the like—on two large screens between the pilots. Although these displays are extremely helpful, there are certain abnormal conditions that the system isn't programmed to recognize, which requires pilots to still have a working knowledge of the aircraft. Every abnormal situation has a checklist, which is to say a list of specific steps to follow, and adherence to checklists (also called "checklist discipline") is critical for flight safety. All pilots are trained in the use of checklists from their first lesson, and departing from them can be perilous. Most checklists on the A330 are computerized and displayed on the computer screens between the pilots, although certain checklists can't be programmed because there are certain conditions the aircraft can't always identify or correctly diagnose, while other conditions are thought to be intuitively obvious or better dealt with in the Quick Reference Handbook (QRH). The QRH is a large and highly detailed set of checklists that is available to the pilots, and it contains more information than the computerized checklists do. Although different airlines and operators use different terminology (e.g., QRH, cockpit operating manual, FCOM, etc.), for our purposes the simple term "checklist" will suffice.

At the time of the accident, one of the things relegated to the paper checklist was the procedure for a fuel leak. A fuel imbalance generated a computer advisory, which is, by definition, something that does not require immediate action, allowing for troubleshooting via the rather detailed checklist. In reality, substantial fuel imbalances are rare, and neither pilot had ever experienced one before. The most common cause of an imbalance in training is flying with a failed engine. In this case, the solution is simple: open a crossfeed valve that connects the left and right systems and turn off the fuel pumps in the tank with lower quantity, allowing the quantities to equalize. This valve is installed on all airliners to prevent the quantities of fuel in the different tanks from varying too much. Because planes store fuel in their wings, having too big of an imbalance can put unintended stress on the plane and make it harder to control. In the case of an engine failure, opening the crossfeed valve makes sense because now the remaining engine won't be consuming fuel only from its own wing. Opening a crossfeed valve in the event of a fuel leak, however, allows fuel that was previously isolated from the leak have access to the leak and can quickly reduce the amount of fuel available to continue the flight.

Practicing engine failures is done in almost every simulator training visit, but training for fuel leaks is not. When presented with the advisory about the fuel imbalance, the Captain opened the crossfeed valve and turned off the pumps in the right wing fuel tank because it had less fuel in it than the left wing fuel tank. Unfortunately, he did this from memory rather than consulting the checklist. Under "Fuel Imbalance," before any checklist steps are listed, there is a caution that reads: "Do not apply this procedure if fuel leak is suspected. Refer to Fuel Leak procedure." Had the Captain not done these steps from rote, the aircraft would have had enough fuel to safely divert to the Lajes airport with fuel to spare (although the right engine would have been shut down). If the Fuel Leak checklist had been consulted, the crew would have read that "The [crossfeed] valve must remain closed to prevent the leak affecting both sides," the exact situation Air Transat 236 was in. While it's true that the most common reason for a fuel imbalance is an engine shutdown, it is also true that any engine failure is an obvious reason for a fuel discrepancy, and unless the reason for the engine shutdown is a fuel leak, crossfeeding is appropriate. In this case, the crew had multiple confusing malfunctions and a more complex scenario. The Captain then opened the crossfeed valve without referring to the checklist (which specifically cautions about a fuel leak scenario). Any time the situation is unfamiliar and more complex than expected, pilots need to slow down and be even more fastidious with checklist discipline. Although there are scenarios that are not specifically outlined in the checklist (it would be impossible to write checklists for every possible permutation of multiple malfunctions), the situation in this case was addressed in cookbook-like form in the aircraft's checklist, reinforcing the need for procedural discipline and adherence to checklists for pilots.

Although it's obvious in retrospect that the crew should have taken the time to consult the checklist, it's also true that the investigation revealed a latent design issue in the A330, namely, the inability of the computer monitoring systems to present unambiguous fuel leak guidance. The investigation revealed two prior incidences of fuel leaks in Airbus aircraft (one each in the A320 and A340, which have similar computer designs). In one case, the crew had no difficulty correctly diagnosing the situation, while in the other, the crew made similar errors to the A330 crew. That professional crews could make this mistake suggested that an improvement in the computer system was in order, and that was subsequently done. Another equally critical measure implemented was enhancing the training of crews in this area. It had previously been thought that this situation would be readily understood by crews, who would then correctly use the checklist to deal with the malfunction, but the reality and the potential severity of the type of error the Air Transat crew made showed that improvements to the system were needed.

Why didn't the crew recognize the problem sooner? The crew started to see symptoms almost an hour before they discovered the imbalance, but neither they nor the Air Transat technical operations specialists (who were contacted by radio) understood the cause of the odd symptoms. Because newer aircraft like the A330 are orders of magnitude more complex than airliners of even 20 years ago, the emphasis has shifted in some areas from detailed systems knowledge to application of computerized and printed checklists. Although these newer aircraft are safer and more reliable than previous generations of airliners, the increased complexity means that sometimes fewer details are memorized about specific valves, circuits, and systems. In years past, frequent examinations would require pilots to know how to draw the entire electrical system on a piece of paper from memory or to trace a theoretical molecule of air from when it entered an engine through when it was discharged at a passenger air conditioning vent. This level of intimate knowledge is generally unattainable given the enormous complexity of newer aircraft, but having a working knowledge of the plane's systems is still valuable.

In this situation, the crew was stymied that the right engine oil temperature was reading 65°C, and the oil pressure was 150 psi. The corresponding values for the left engine were 110°C and 80 psi. These readings were bewildering because both engines had been reading approximately the same earlier in the flight. What would cause them to diverge? To many pilots raised on earlier jet airliners (DC-9, B-727, etc.), these values pointed to a vastly increased flow of (very cold) fuel across the fuel/oil heat exchanger (cold fuel is used to cool hot oil). A prime suspect with those symptoms would be a fuel leak. This crew was experienced, but perhaps because of the insularity provided by the computerized monitoring systems they were unable to correlate these symptoms with any particular system malfunction. More puzzling, the dearth of insight into the situation also existed in the Air Transat operations center, where maintenance personnel were unaware of the implications of these readings. As pilots, it's important when something changes to discover the cause. In this case, the crew attempted to do just that but found the symptoms difficult to diagnose even with the assistance of specialists and onboard manuals. While they were troubleshooting this issue, the fuel imbalance was increasing, but they were using the lower computer display screen to examine the "Engine" status page, which prevented the fuel system from being displayed in detail. Further masking the problem was the functioning of a fuel tank (the "trim tank") in the tail of the aircraft that was operating in an automatic mode. In normal operations, when the fuel in the wing tanks got to a predetermined low level (4 tons, near the conclusion of the flight) the trim tank would transfer its fuel to the wing tanks, evenly divided between the right and left wing tanks; however, with a significant imbalance, all the fuel would be transferred into the low tank. In this case, the right tank was lower because of the massive leak, so this resulted in even more fuel being inadvertently dumped overboard, but somewhat deceptively kept the fuel balance more reasonable on cursory examination between the wing tanks for an additional 15 minutes.

Although the worsening leak went undetected while the crew tried to reconcile their understanding of the confusing picture the displays presented them, when they did address the fuel status, they greeted it with disbelief. Although it had been going on for 55 minutes, when the crew finally recognized there was a problem with the fuel system, they thought an unknown computer problem was giving them improbably erroneous indications. They couldn't come to terms with a scenario that accounted for all the discrepancies they were facing. They also had not been exposed to this scenario in training, nor had either ever experienced an imbalance of this magnitude in their years of flying. The crew recognized quickly that there was a shortfall of fuel on board and that there was an indicated imbalance, but they did not recognize the magnitude and significance of the actual fuel depletion until it was too late. Even though the modern computerized planes are significantly safer, this is one of the few examples where automation confused instead of helped the pilot.

Diversion to Lajes

There are few scenarios more terrifying to a professional pilot than ditching an aircraft in an ocean at night, but after the engines stopped running, the aircraft became a 168-ton glider. Mercifully and by sheer luck, the aircraft had been routed on a fairly southern route across the Atlantic Ocean that evening, as a track farther north (which was and is common) would have put the A330 out of gliding range of Lajes; likewise, if the aircraft had not had the extra 12,100 lbs of tankered fuel on board when departing Toronto, the plane would have been unable to reach the island airport. Once the crew realized that the aircraft had too little fuel onboard to safely complete its flight, they began an immediate diversion to the nearest suitable airport, and from that point onward, the crew performed remarkably well. The right engine failed about 150 miles from Lajes, and the left failed shortly thereafter, 65 miles from the field. At that point, the Captain manually flew the aircraft with severely restricted emergency instruments and with degraded systems, while the First Officer supported him by executing procedures, checklists, making public address announcements, and so on. When the second engine failed, three things quickly became imperative: flying directly to the airport (in this case, it was already being done), preparing the cabin for an emergency ditching or landing (about which the Captain had coordinated with the Flight Attendants), and flying a precise speed that yields the maximum range. This speed (which varies substantially with weight and is normally substantially slower than cruising speed) is provided by the cockpit computer and provides the best lift-to-drag ratio, referred to by pilots and engineers as L/D_{MAX}. The report found that the Captain's handling of the aircraft during the glide and subsequent landing "were remarkable given the facts that the situation was stressful, it was night time, there were few instruments available, pitch control was limited, and he had never received training for this type of flight profile." By flying the aircraft accurately, the crew arrived with more altitude and airspeed than needed or wanted at Lajes. This is the ultimate pilot's dilemma: it is imperative to glide far enough to reach the island (using years of experience instead of sophisticated instruments and computers, which don't work in this scenario), but when arriving, if there is excess energy (i.e., too much altitude or airspeed), it is difficult to get rid of. Ultimately, there is only one chance to land on the runway when gliding, and if excess speed is present, it will have to be dealt with by the aircraft's braking system. In this case, the aircraft bounced once and had a firm second touchdown: numerous tires burst owing to excess brake energy, and small fires occurred at the left landing gear (which were quickly extinguished by the fire department).

Despite these issues, on balance, the aviation community believes that the crew performed magnificently during the glide and landing. Some find it ironic that it was when the crew reverted to manually flying the aircraft that they performed with admirable skill, doubtlessly saving the lives of all aboard: it wasn't the flying that proved difficult for them; it was understanding the "big picture" story from the indications that their automated aircraft was giving them.

Lessons Learned

After the Air Transat 236 incident, several changes were made to further enhance safety. First and most obviously, increased training emphasis is now placed on understanding fuel leaks and fuel crossfeeding procedures, and aircraft manuals were in some cases revised to make them more user-friendly in confusing fuel imbalance scenarios; likewise, checklist discipline in general is more emphasized since the Air Transat incident. Though the most common cause of an imbalance in training is still an engine failure, other scenarios are now covered more thoroughly in most countries, and many airlines have extensive and complex fuel leak simulator training scenarios. People wondering why fuel leak scenarios are not taught at every simulator training session must understand that training focuses on the most likely scenarios that are perilous to the aircraft and passengers, and fuel leaks of this magnitude are rare. Some situations are required to be reviewed every time (engine failures, etc.), while others are rotated over a series of training cycles. Fuel leak scenarios take a long time to set up and teach in the simulator, and simulator time is a scarce resource. Most airlines and regulatory agencies generally teach things far more likely to occur, as that is where the greatest overall safety gains can be made.

Most initial schools have exposure to fuel leaks of different varieties now, and this training recurs with a degree of regularity. Overall, this is a far cry from the situation prior to the incident.

Maintenance changes were required, and now configuration control of various spare parts and maintenance procedures were substantially revised in light of the incident: Air Transat was audited by Transport Canada (the Canadian equivalent of the Federal Aviation Administration) and received the largest fine ever levied on a Canadian airline for maintenance errors up to that time. Finally, regulatory bodies including the FAA required low-fuel warnings to be improved on all long-range aircraft flying over oceans.

8

Approach

After an 11-hour flight from London, British Airways Flight 38 landed in Beijing with about 17,860 lbs of fuel. After loading 157,400 lbs of fuel, the Boeing 777 took off on January 17, 2008, for the 13-hour return flight. Shortly after takeoff, air traffic control assigned an altitude of 34,800 feet. Because of predicted severe cold, the flight crew planned on monitoring fuel temperatures.

Because of fuel burn and weight reduction, without a power or trim adjustment, a plane would tend to climb gradually on its own. Airlines generally prefer to operate at the highest altitude in the thinnest air to reduce drag and fuel usage. Air traffic controllers, however, must know the plane's altitude with precision. In practice, airplanes in cruise flight almost always have the autopilot engaged, which holds the altitude and speed constant. The autopilot adjusts flight control surfaces, including trimming the horizontal stabilizer as required to maintain speed at a given power setting; the autopilot is integrated with the autothrottle system, which adjusts engine thrust to maintain desired aircraft speed. Planes are authorized by air traffic control to make altitude changes in steps during cruise. About 350 miles north of Moscow, the plane climbed to 38,000 feet. Over Sweden, the 777 climbed again to 40,000 feet. For passenger comfort, the step changes were accomplished using the vertical speed mode of the autoflight system to minimize engine thrust and noise.

The flight continued uneventfully until the later stages of final approach. At 1,000 feet and 83 seconds before landing, the landing gear was down and Flaps 30 was selected (see the Boeing 777 flaps in Figure 8.1), with the autopilot and autothrottle engaged.

Figure 8.1. Flaps and landing gear extended on a Boeing 777 seconds before landing. Wikimedia/Arpingstone

Flaps 30 is the highest of six flap settings, with flaps deflected downward at 30°, the maximum amount, to slow the plane and create more lift at lower speeds. The First Officer took control of the aircraft at 800 feet and intended to disconnect the autopilot at an altitude of 600 feet. Shortly after the First Officer took control, the autothrottle commanded the first of four thrust increases in both engines.

Airspeed gradually reduced to the target approach speed of 135 knots. During the fourth thrust increase, the right engine (altitude 720 feet) followed by the left engine (altitude 620 feet) experienced a reduction of thrust. At a height of 430 feet (34 seconds before touchdown, airspeed of 133 knots), the Captain announced that the approach was stable, and the First Officer responded "just." The crew was beginning to understand something was amiss. Seven seconds later, the First Officer asked, "What's happened to the speed . . . What's going on?" The pilots quickly analyzed the situation as they received "Master Caution" alert lights on the glareshield in front of

Why Can't the Plane Glide to a Safe Landing?

Last-minute thrust is needed because of changes in winds and because the plane is flying with significant drag, with full flaps (in this case 30°) and landing gear down. The bottom line is that it takes a lot of thrust to fly slowly with flaps out and landing gear down. Planes typically descend in idle thrust, as the engines are essentially turning as slow as they can while remaining lit and stable and powering their accessories (pumps, generators, etc.). For all intents and purposes, in an idle descent the aircraft is effectively gliding (the small amount of residual thrust involved can be safely ignored). Pilots use throttle near the ground for two reasons: (1) because it requires a lot of power to make a controlled descent at about 3° on a normal approach fully configured and (2) you do not want to be in idle below 1,000 feet above the ground because there is an inherent "spool-up" time for engines to attain high power from idle (this lag is on the order of four to eight seconds, depending on the engines) in the event of a go-around or wind shear, being in idle close to the ground has caused a number of serious accidents.

them along with an "Airspeed Low" message accompanied by an aural alert consisting of four electronic beeps.

The Captain, attempting to reduce drag, reduced the flaps from 30 to 25. Boeing analyzed this maneuver and estimated that it extended the touchdown by 164 feet, allowing the plane to miss the instrument landing system antenna. British investigators concluded that this quick thinking prevented substantial fuselage damage, but there was a trade-off, as the lower flap setting also decreased lift.

The airspeed continued to reduce and was 115 knots at 240 feet and 108 knots at 200 feet. Ten seconds before impact, the stick shaker activated, indicating an imminent stall. The autopilot was still engaged and was trying to track the 3° glideslope of the approach by raising the nose of the aircraft and in the process was compensating for the slowing speed by increasing the angle of attack (AOA). The autopilot disconnected when the First Officer pushed the control column forward before pulling back immediately before impact. The plane struck the ground about 1,080 feet short of the runway with a sink rate of 25 ft/sec (1,500 ft/min). The reason the First Officer pulled back prior to ground contact was to minimize the rate of descent and impact damage, which would have been even greater had the high sink rate not been reduced.

Recall that lift is proportional to velocity squared. If the plane tries to fly straight and level with decreasing speed, the pilot must increase the AOA to create more lift. Eventually the AOA becomes excessive, and the airflow does not follow the contour of the wing and instead "separates" at the critical AOA—the definition of an aerodynamic stall. Stalling can be abrupt and is extremely dangerous, especially so when close to the ground.

If the plane instead tries to fly straight and level with constant AOA at a speed not quite high enough to create a lift force that balances the plane's weight, an additional force is required to create equilibrium. The plane's vertical sink rate increases until a vertical drag force plus the wing's insufficient lift balances the plane's weight. The plane is now sinking at a constant speed, much like a skydiver at terminal speed. If the plane's forward speed continues to decrease and reduce wing lift (assuming constant AOA), the plane now reaches equilibrium at a higher vertical sink rate. Increasing sink rate is a gradual process that is normally easily corrected with increased engine thrust. An increased sink rate is problematic if too close to the ground and was especially problematic for Flight 38. Flight 38's engines failed to produce additional thrust when needed.

Flying the plane at too low a speed is not stalling per se. Stalling is a term reserved for exceeding the critical AOA. The autopilot on the accident flight was tracking the instrument landing system glideslope, so as speed decreased, it raised the nose higher and higher, with a net effect of increasing the AOA. As the speed slows and the angle of attack increases, an aerodynamic stall eventually occurs.

Captain Hedges Explains: Can the Autopilot Land the Plane?

In most modern airliners, the autopilot can be programmed to land the aircraft; however, it is rarely done because pilots normally do a better job and pilots need to retain proficiency, especially on long-range flying like the B-777 does, as pilots sometimes only get one landing per month owing to crew scheduling (in the United States, for example, pilots must get at least three landings every 90 days).

Do Only Fly-by-Wire Planes Have Autoland?

Both fly-by-wire (FBW) and non-FBW aircraft have autoland, and typically they are done for low-visibility conditions—that is, below about a half mile of visibility—though some operations can be hand-flown at lower visibility. Autoland systems have strict limitations—for example, maximum headwind, crosswind, and tailwind limits are specified as are the acceptable flap configurations and other similar details—so it can't always be used, and each type of plane's limitations are slightly different.

Do the Pilots Take Turns Landing for Practice?

Normally the Captain and First Officer alternate flights (or "legs"), but note that on Flight 38 there were three pilots, so on the two-leg trip, one pilot didn't get a landing, which is why the currency issue mentioned before is so critical and autolands in good weather are relatively rare, particularly on long-range aircraft.

Preceding aerodynamic stall is low-speed buffet (i.e., boundary-layer flow separation described in Chapter 4) that is turbulent air flowing off the wings and affecting the tail. This turbulent air shakes the tail section. The vibrations pass through the controls and shake the control column. The effect is artificially created in some modern jets (including the Boeing 777) with a stick shaker to warn the pilot of impending aerodynamic stall. The plane's estimated weight at the time of the accident was 410,880 lbs. (The plane used nearly 150,000 lbs of fuel flying from Beijing.) Boeing calculated the stall speeds for this weight to be 104 knots at Flaps 30 and 106 knots at Flaps 25. Boeing's analysis, for the conditions of Flight 38, predicted stick shaker activation at 4 knots above the stall speed.

The plane did not fall like a rock but instead descended rapidly because there wasn't enough lift to properly balance the weight of the plane on the desired descent profile. As the sink rate increased so did the vertical component of drag, a helpful effect. Unfortunately, the aerodynamic lift was decreasing too fast to allow for landing at a normal descent rate.

The plane hit about 1,082 feet short of the concrete runway at a vertical speed of 25 ft/sec (17 mph vertically) and with a recorded impact of 2.9 G. Impact at 25 ft/sec is equivalent to falling like a rock from a height of only 9.7 feet. By way of comparison, a section of Boeing 787 fuselage passed a drop test with impact at 30 ft/sec. The 787 fuselage section was dropped from nearly 15 feet without the benefit of attached landing gear. (Passing means the crash test dummies are injury-free. Crash test dummy criteria are described in *Beyond the Black Box*.)

The touchdown dug holes in the soft soil about 18 inches and 14 inches deep under the right and left landing gear, respectively. The rear fuselage also made ground contact as the pilot was trying to arrest the descent rate by attempting to flare (i.e., raise the nose) with the last-minute pull on the control yoke. After first contact, the plane bounced and became airborne for about 174 feet. The main landing gear, engine nacelles, and nose landing gear all left a ground scar during the second impact. During the ground slide, the plane was greatly slowed by both engines scooping up soft soil. The plane slid for over 1,000 feet and came to a final stop 1,220 feet after first impact. Security cameras documented the time line given in Table 8.1.

An example of a "normal" instrument landing approach with a 3° glide path is shown in Figure 8.2. Flight 38 was on a 3° glide path, as is published on the instrument landing system chart for Heathrow's runway 27L. Flight 38 was about 2.6 miles from the runway threshold when the problems begin at an altitude of 720 feet above ground level (AGL). The plane's first touchdown was almost 1,700 feet short of the "perfect" 3° landing.

Preoccupied with flying the plane, their primary responsibility, the flight crew was unable to warn the passengers. A "Mayday" call was transmitted to controllers three seconds before touchdown. British newspapers alter-

Table 8.1. Sequence of Events

Time	Event
12:42:09	First impact
	Plane becomes airborne, second impact, and slide out
12:42:22	Controllers contact fire crew
12:42:51	Doors begin to open
12:43:07	Passengers descend slides
12:44:13	First fire trucks arrive
12:45:11	Empty slides on right side
12:55	Firefighters enter cabin

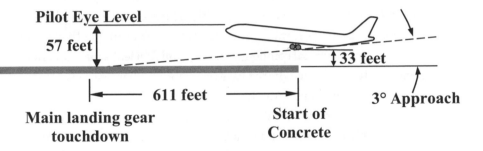

Figure 8.2. Example glide path for a 3° approach. Main landing gear is 33 feet above the start of the paved runway.

nately described the passengers as calm or screaming in terror, fearing they would die as they ran through a smoke-filled plane. A reporter, who witnessed the event on the runway in a nearby plane, wrote: "a great cloud of black smoke, flame and flying bits of blackness."[1] The official report contradicted the reporter. There was no smoke or flames: no flames in the engines, and no flames from spilt fuel occurred. Presumably, the slide across the dirt created a cloud of dust. This reinforces the conflicting testimony eyewitnesses can provide in response to an unexpected and traumatic event.

Fleeing passengers feared the wounded plane would explode. It turns out that nothing burns except flammable gases. Paper and even gasoline do not actually burn per se. Hot paper decomposes into flammable gases. Gasoline vapors are dangerous, but a match submerged in gasoline will not ignite. Jet fuel spilt on the ground will not explode, although jet fuel vapors can explode if contained inside the fuselage. Containment is required for combustion gases to create a dangerous pressure buildup. Ignition will only occur with temperatures above the flash point. Flash point is the lowest temperature at which jet fuel will evaporate a sufficient amount of liquid to create an ignitable mixture in air. The flash point for Jet A-1 fuel is at least 100°F; realistically, it is probably 10°F to 20°F higher. Jet fuel can explode if contained inside the cabin (it was not), the temperature is above 100°F, and there is an ignition source.[2] Gasoline vapors, with a flash point of –40°F, are very dangerous. In aviation accidents, crews are trained to get passengers out of the aircraft and away from it quickly, as fires can be unpredictable: there are almost always significant ignition sources available (hot engine components, live electrical components, etc.), and fuel leaks may not always be immediately recognized. In the case of Flight 38, there were significant fuel and oxygen leaks, a potentially serious hazard.

Plastics inside the cabin will burn, and their decomposition products can be extremely toxic. All airliners are designed and certified for complete evacuation in less than 90 seconds with half the exits blocked. With significantly safer planes and more fire-resistant materials, there have been only two high-fatality fire events in the developed world since the 1980s.[3]

Landing Gear

Landing gear (a.k.a. shock strut) is designed to absorb the plane's kinetic energy of motion in the downward direction—within design limits. Two cylinders are used. The upper cylinder is attached to the fuselage and does not move. The lower cylinder slides in and out (i.e., telescopes) into the upper cylinder. The upper cylinder is filled with nitrogen; the lower cylinder is filled with hydraulic fluid. Motion of the lower cylinder forces the hydraulic fluid through an orifice from the lower cylinder into the upper cylinder. More hydraulic fluid pressure will force the hydraulic fluid through the orifice at a faster rate, allowing the lower cylinder to telescope faster into the upper cylinder. Allowing flexibility in the design, the resisting force depends on two different effects. The resisting force of the compressed nitrogen depends on the amount of compression; the resisting force of the hydraulic flow through the orifice depends on the sink rate. A higher sink rate automatically creates a higher resisting force. A simplistic schematic is shown in Figure 8.3.

The lower cylinder does not directly attach to the wheel as shown, but instead attaches to a truck assembly that mounts the axles and wheels. There are additional links and braces to prevent the lower cylinder from falling out, to extend and retract the gear, and to lock it in place for landing.

When first deployed for landing, the lower cylinder is fully extended. Additional flow paths and chambers cushion the shock of extension and rebound. Hard metal-to-metal contact during full extension of the heavy steel parts without hydraulic cushioning would be loud and disturbing. The pressure required to force the hydraulic fluid through the orifice is further refined by a tapered metering pin attached to the top of the lower cylin-

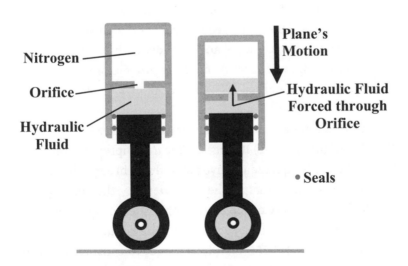

Figure 8.3. Landing gear. The lower cylinder telescopes into upper cylinder.

der. The metering pin moves through the orifice and alters the flow area of the hole. The metering pin allows additional flexibility to custom design a changing force with changing sink rate. When on the ground, the compressed nitrogen supports the weight of the plane.

Both of Flight 38's main landing gear partially separated during the initial impact. During the subsequent ground roll, the left main landing gear collapsed (i.e., folded under the plane) and the right gear separated. Main landing gears are attached with 12 precision mechanical fuse pins designed to control the breakaway sequence and protect the fuel tanks in the wings. The separation of the left main landing gear, followed by the proper breakaway sequence, protected the fuel tanks and remained partially attached. Sideways forces and complex interaction with the soft soil prevented the right main landing gear from following the breakaway sequence. The right main landing gear broke off and punctured the fuselage (breaking a passenger's leg, the only serious injury) and ruptured a fuel tank. Fortunately, the spilt fuel was not ignited. (With expected passenger evacuation under 90 seconds, ignited fuel most likely is not a catastrophe.) With the plane pitched up, the nose gear did not contact the ground until the second impact, upon which it promptly collapsed.

Landing gear is mandated to withstand impact up to 10 ft/sec without permanent damage to the plane or the gear. A crushable element inside the landing gear absorbs additional energy and provides additional protection up to 12 ft/sec. Because impact kinetic energy is proportional to velocity squared, the landing gear absorbed $(12/25)^2 = 0.23$, or 23% of the impact energy.

A physicist defines work as force × displacement. (It takes 10 foot-lbs of work to lift a 5 lb weight up 2 feet.) It takes a force to compress the nitrogen and an additional force to squeeze the hydraulic fluid through the orifice. Applying the conservation of energy, the work done by the landing gear must absorb the kinetic energy of the plane's sink rate. In equation form: force × distance = ½ × mass × velocity². Assuming the landing gear strokes 2.2 feet, the plane weighs 410,880 lbs (don't forget to convert weight to mass by dividing by the acceleration of gravity = 32.2 ft/sec²), and using the design maximum sink rate of 10 ft/sec, the average force supplied by both landing gear when landing at a sink rate of 10 ft/sec is 290,000 lbs. The force on one gear is half, or 145,000 lbs. This is an average force; the actual maximum force will vary in complex ways depending on the actual flow rate of hydraulic fluid through the orifice.

The Investigation

The investigation quickly zeroed in on a sudden power loss in both engines. Unlike unreliable piston engines, there has never been a double engine failure from two different mechanical problems in the modern jet era. Multiple jet engines have occasionally failed from a single problem (i.e., fuel exhaustion, volcanic ash, etc.). The problem had to be something common to both engines. More specifically, fuel flow rates recorded by the flight data recorder were abnormally low during final approach.

To reduce noise and wear and tear, planes normally take off with the least possible thrust that meets all the mandated safety requirements defined in Chapter 2. Maximum flow rates for the two engines at takeoff were 24,176 and 23,334 lbs/hr. Climb was completed with peak flow rates of 8,896 and 8,704 lbs/hr.

Fuel flow did not exceed 1,250 lbs/hr during the first six minutes of descent—the plane was essentially gliding. Gravity during the descent provides the power to maintain speed. The plane leveled off before entering a holding pattern, and fuel flow increased to about 4,900 lbs/hr. After extending flaps and lowering the landing gear, fuel flow increased to 7,300 lbs/hr. Because of speed and drag changes, the autothrottle commanded a series of four thrust increases. The second thrust command resulted in the highest fuel flow of 12,288 and 12,032 lbs/hr in the left and right engines, respectively. During the fourth autothrottle thrust increase, the fuel flow in the left and right engines increased to 11,056 and 8,300 lbs/hr, respectively, before gradually reducing. The left engine stabilized at about 5,000 lbs/hr and the right engine at about 6,000 lbs/hr. About 20 seconds from touchdown, the autothrottle and the flight crew applied full throttle (greater than 38,000 lbs/hr) without any recorded increase in fuel flow; the engines did not receive enough fuel to maintain safe flight.

Also significant was the low fuel flow during cruise. Average fuel flow during cruise from the main wing fuel tanks for each engine was just 6,900 and 6,800 lbs/hr. Mindful of disturbing the passengers during this nighttime flight, the flight crew selected gentler climbs using the vertical speed autopilot mode instead of using vertical navigation (VNAV), a more automated mode but one that would be more abrupt to the passengers by using more thrust and a higher climb rate. The maximum fuel flow rates for the two step climbs during cruise were 8,896 and 8,704 lbs/hr.

After eventually ruling out electrical/computer/software issues, the investigators began to focus on, of all things, fuel line freeze-up. Everyone expects the black box data to solve all accident mysteries. With over 1,400 recorded parameters, eyewitness reports from the flight crew, and a mostly intact plane, a quick resolution was expected. Instead, the investigation took over two years. The primary evidence had melted!

Water

A small amount of water normally exists in jet fuel. The fuel was tested and found to contain 35–40 parts per million of water, well within established limits. This amount converts to about 1.36 gallons or 314 cubic inches of water. The water can be dissolved or free. Dissolved water molecules are chemically bonded to fuel molecules and do not form ice. As the fuel temperature decreases, dissolved water comes out of solution becoming free water. Free water behavior depends on the size of the water droplets.

Being denser than fuel, large water droplets settle on the bottom of the tank. Because a large plug of water could affect engine stability (or lead to microbial growth), the water on the bottom of the tank must be regularly drained. Routine maintenance was done on the plane before it departed for Beijing on January 15, including a fuel sump drain. British Airways procedures for draining water were reviewed and found appropriate. Additionally, all British Airways Boeing 777s were inspected for excessive water buildup, and none was found. To prevent water buildup, the Boeing 777 also has a water scavenge system with additional pumps that continuously draw fluid from the lowest part of the fuel tank. Free water on the bottom of the tank freezes into hard and stable ice that stays put. The scavenger system and sumping were not considered factors in this accident.

Water droplets too small to settle to the bottom of the tank become suspended or entrained water. Suspended water at concentrations below 30 parts per million is clear; at higher concentrations, the fuel becomes cloudy.

Entrained water can settle with time or grow larger by linking with other entrained droplets. Entrained water will flow with the fuel. Entrained or suspended water will freeze and form ice crystals.

The transition of dissolved water to suspended water is expected on long flights. The ice crystals begin to form at 27°F to 30°F. Because the density of ice crystals is nearly the same as the fuel, the ice drifts with the fuel until contacting a cold surface. As the temperature is lowered, the crystals will stick to cold surfaces; at lower temperatures still, the ice crystals will stick to each other. (Soft ice formation on the fuel feed tubes is the most likely scenario for Flight 38.) The temperatures at which ice crystals stick to surfaces or to each other is sometimes called the sticky range and was identified experimentally to be −4°F to 23°F. Worse still, ice in the sticky range is soft and easily dislodged during the higher fuel flows associated with final approach.

Fuel Temperature

At takeoff, the fuel temperature was 28°F. At 38,000 feet between the Ural Mountains and eastern Scandinavia, the plane passed through a region of particularly cold air, −101°F. All petroleum-based fuels consist of mixtures with different molecular-length hydrocarbon molecules and various trace contaminants that vary with each individual oil well and refinery. For this reason, there is no definitive specification for the chemistry of fuel. Instead, jet fuel must meet 30 performance specifications, including a defined freezing point no higher than −53°F. Properties, including freezing point, will vary slightly from batch to batch. The fuel on this flight was tested and froze at −71°F, greatly exceeding the mandated specification. The fuel does not freeze per se, but forms wax. About 7.5 hours into the flight, the fuel temperature reduced to its lowest level of −29°F and remained there for the next 80 minutes. (Captain Hedges adds: Pilots monitor fuel temperature, particularly on long-range and polar flights, and if the fuel temperature approaches a low-temperature limit, the two basic choices available are to descend into warmer air or to speed up, as higher aircraft speed increases the friction of the air passing over the wing and measurably heats the wing and the fuel inside the wing fuel tanks.)

The Fuel System

The fuel in a B-777 is contained in three fuel tanks—one in each wing and a center fuel tank between the wings. Each fuel tank contains two fuel pumps supplying fuel to the engine fuel system.

The pressure ratio for the Rolls-Royce Trent engines is 42; that is, the outlet pressure is 42 times higher than the inlet pressure. To overcome the pressure drop for exhaust gas blowing through the turbine section, the combustion chamber must operate at an even higher pressure. To overcome the pressure, atomize the fuel, and provide mixing with combustion air, the fuel is pumped in at high pressures—typical values are 1,100 to 1,900 psi. This pressure is supplied by the engine fuel system high-pressure pump (not to be confused with the pumps in the fuel tanks).

All pumps create a higher pressure at the outlet and a lower pressure at the inlet or suction side. If the inlet pressure is too low, the fuel can flash to vapor, creating fuel vapor bubbles. Boiling water experiences similar changes with pressure. If higher pressure in a sealed pressure cooker raises the boiling temperature, then, conversely, lower pressure on top of a mountain lowers the boiling temperature.[4]

There is a second effect also trying to create bubbles in the fuel. As the pressure is lowered at the pump inlet, dissolved air molecules will come out of solution, creating air bubbles. The effect is like opening a can of soda nominally pressurized at about 50 psi. Opening the can reduces the pressure to 0, and the dissolved CO_2 comes out of solution in a familiar fizz.

The tiny bubbles (filled with air and fuel vapor) pass through the pump to the high-pressure side and collapse or implode. The implosions send out pressure shockwaves of many thousands of pounds per square inch in a process known as cavitation. Pressure that high would normally destroy most containers. The bubbles, however, are very small. Cavitation, a well-known problem of pump design, will damage the pump housing with surface erosion. The problem is so severe that an additional pump is needed to maintain sufficient pressure on the suction side of the high-pressure fuel pump. The high-pressure fuel pumps for both engines showed fresh cavitation damage.

Cavitation can also be caused by a flow restriction upstream from the inlet. It's as if an orifice plate is placed in the flow stream. The orifice creates increased pressure upstream of the orifice hole and reduced pressure downstream. There is a pressure drop across the hole caused by the churning turbulence of fluid trying to pass through the restriction. Here's another way to look at cavitation, with a restriction: if the outlet is trying to pump 10,000 lbs/hr and the restriction allows only 8,000 lbs/hr to make it to the pump inlet, the pump tears a void in the fluid—a void that appears as bubbles. Icing tests of the fuel system could only duplicate a restriction at the oil/fuel heat exchanger.

An extremely simplified schematic of the fuel system is shown in Figure 8.4. Also shown are the fuel/oil heat exchanger and the fuel metering unit. Not shown are several filters and various valves (to isolate any leaks or bypass plugged components).

The Rolls-Royce Trent 800 series engine has three rotating shafts. The shafts are mounted on lubricated bearings. The high rotational speeds generate heat in the oil. Ice can form in fuel at low temperatures. This ice can affect the delicate computer-controlled fuel metering unit. To remove the ice, the heated oil is used to warm the colder fuel in the fuel/oil heat exchanger (FOHE).

The fuel enters the top of the FOHE and passes downward through 1,180 small tubes (about 0.08 inches in diameter) that protrude about 0.16 inches above the inlet face. (After the accident, it was discovered that the protrusions collected soft ice.) The cold fuel enters the larger cylinder and flows around the outside of the tubes filled with hot oil. This arrangement is commonly called a shell and tube heat exchanger and is shown in Figure 8.5.

Boeing and Rolls-Royce conducted significant cold fuel testing. It was determined that 1.5 cubic inches of water could restrict fuel flow in the fuel/oil heat exchanger. (Blockage elsewhere in the system was ruled out by a variety of means.) Investigators did not use the word "snow," preferring instead the phrase "soft ice." Nor did they specify the density of the soft ice formed during testing. Possible densities of snow may give insight. When water freezes, it can form a variety of different crystal structures (i.e., wet and dry snow and everything in between). Volume of snow can vary between 8 to 22 times greater than the water content.

Regulations require the fuel system to work with a water content of 200 parts per million, significantly higher than the measured content for the ac-

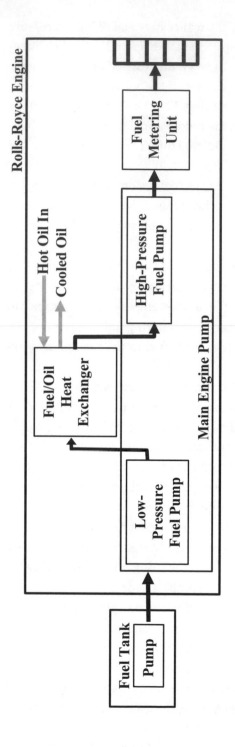

Figure 8.4. Simplified schematic of fuel system.

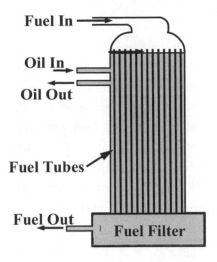

Figure 8.5. The fuel/oil shell and tube heat exchanger.

cident flight. Also, the plane was certified for flight at −103°F. But none of the certification tests considered the circumstances of this flight: the possibility of operations at extremely cold temperature, low fuel flow during cruise (allowing soft ice to accumulate on surfaces), and higher-than-normal fuel flow during approach (which knocks off the ice).

Data Mining

Data mining is a process derived from statistics, artificial intelligence, and machine learning. Data mining techniques are used to analyze large amounts of data to identify patterns. The methodology has been developed and used over a period of years in both science and business. A team of statisticians—together with specialists from the Boeing, Rolls-Royce, British Airways, and the British crash investigators—conducted a review of data from the accident flight and from other data sources. Data for 175,000 Boeing 777 flights were entered into a computer. Through data mining, investigators learned that Flight 38 had a unique combination of low fuel temperature, low fuel flow during cruise, and high fuel flow during approach.

The Fix

Boeing issued new extreme cold operating procedures. These included, among other things, pilots using maximum engine thrust within three hours prior to descent to interrupt and blow out any accumulated ice. If reduced thrust still occurs, tests showed that reducing thrust to idle would melt any accumulated ice on the fuel/oil heat exchanger. Reducing the flow of colder fuel through the heat exchanger results in less cooling of the hot oil, and hotter oil flowing through the heat exchanger melts accumulated ice. Reduced fuel flow also reduces the ice flowing into the heat exchanger. If reduced thrust occurs, the new Boeing procedure called for descending to a lower altitude while reducing engine thrust to idle.

Fuel line freeze-up struck again on another Boeing 777 with Rolls-Royce engines in November 2008. In this case, one engine stalled at 39,000 feet. The flight crew was uncertain whether the new procedures applied. The thrust reduction persisted for 23 minutes until descending to 31,000 feet. In the November 2008 incident, pump cavitation damage was similar to that on Flight 38. Boeing again revised their cold fuel procedures to require reducing engine thrust to idle sooner during the descent to reduce the time of ice buildup.

Why didn't General Electric (GE) or Pratt & Whitney engines seem to have an icing problem? There are two outlets from the flow metering unit (FMU): one flow stream to the burners and another flow recycled back to the fuel system. The purpose of the recycled flow is to provide more precise computer controlled flow metering than can be accomplished by adjusting pump power. Fuel exiting the FMU has been heated by passing through the fuel/oil heat exchanger and the high-pressure fuel pump. GE and Pratt & Whitney recycled the warmer fuel from the FMU to a point downstream of the fuel/oil heat exchanger, whereas Roll-Royce recycled the fuel from the FMU upstream of the heat exchanger. The GE and Pratt & Whitney arrangement provided a little more heat to fuel entering the fuel/oil heat exchanger than the Rolls-Royce design.

Mindful that additional procedures can result in operator error, Rolls-Royce devised a permanent fix. The tubes in the Rolls-Royce heat exchanger protruded about 0.16 inches above the face (presumably to increase turbulence and heat transfer). Unfortunately, the protrusions also collect ice. The new Rolls-Royce heat exchanger design removes the tube extensions; all the tubes are now flush with the end plate. Rolls-Royce has successfully completed icing tests with fuel 18°F colder and with 5% more water content than the accident flight. Installation of the new "drop-in" design was completed by September 2010.

Captain Hedges Explains: Airmanship and Dealing with the Unexpected

Air Transat 236 (discussed in Chapter 7) demonstrated that even large and heavy aircraft can glide successfully for long distances given enough altitude and airspeed, even without fuel. The advantage the Air Transat pilots had in preparing for their glide was an awareness of the fuel leak that made it inevitable, which gave the crew adequate time to plan and execute the glide to a successful landing in the Azores. At the other end of the spectrum is the accident involving British Airways Flight 38 on January 17, 2008, when the crew found themselves suddenly running out of airspeed and altitude just 48 seconds before touchdown.

Long-range aircraft like the Boeing 777 frequently fly for 10 hours or more. Airlines try to minimize flying time for economy, and on long-range flights to and from destinations in the Northern Hemisphere, this frequently results in flying at extreme latitudes near the North Pole. In the case of Flight 38, after departing Beijing, the aircraft headed up over Mongolia, Siberia, and Scandinavia, where the temperature outside the plane was as low as -74°C (-101°F), a temperature officially termed "unusually low compared to the average but not exceptional." Pilots flying long international routes must monitor fuel temperature, as these aircraft have a minimum fuel temperature limitation depending on the type of fuel used; for Flight 38, the minimum fuel temperature reached was -34°C (-29°F). The type of fuel used (Jet A-1) freezes at -57°C (-70°F). If fuel gets too cold, it can get waxy and will not flow correctly. If the minimum fuel temperature is reached, pilots will normally need to descend to warmer air, though in some instances speeding up the aircraft will heat the wing sufficiently owing to the friction of the more rapid airflow. (Like most airliners, the B-777 carries most of its fuel in its wings.) There was no need to alter the flight profile of British Airways Flight 38, and in fact during the flight the aircraft performed climbs to higher altitudes. Such "step climbs" are common on long flights, as they help minimize fuel burn; as the aircraft burns fuel, it gets lighter and can climb to a higher, more efficient altitude. Modern aircraft are automated, but there is still much technique used by pilots to optimize the flight. One common technique used for climbing the aircraft a small amount (like the step climbs made by Flight 38) is to select a small vertical speed climb, which is not nearly as abrupt as going to maximum climb power (which is what the highest level of vertical automation—vertical navigation, or VNAV—would do), especially at a tranquil point during cruise while most passengers are asleep. Pilots strive to provide the smoothest, most comfortable flights possible to the passengers, and this climb method was considered good technique. One side effect of this method is that it doesn't raise the fuel flow through the engines nearly as much as a full power climb, and in the case of Flight 38, the fuel flow through the engines was quite low compared to other flights of similar durations. Although that was great from an efficiency viewpoint, it had unexpected consequences later.

As British Airways Flight 38 continued its journey to London, all seemed well, and after a quick turn in a holding pattern, they were vectored to intercept the instrument landing system (ILS) approach to runway 27L at Heathrow Airport. As mentioned before, the First Officer was to be the pilot flying the approach, and he had the autopilot and autothrottles set to fly the airplane on the correct course, descent angle, and speed to the runway. As the plane slowed, the crew correctly configured the aircraft with the landing gear down and the flaps to 30°. It may seem counterintuitive because the aircraft is flying at a much slower speed on final approach than at any other point during the flight, but the final approach segment requires a large amount of power and much higher fuel flows than during the cruise portion of the flight. To put these fuel flows in perspective, each engine burned about 24,000 lbs/hr during takeoff, while during cruise each engine burned about 6,900 lbs/hr. When configured at final approach speed and with the landing gear and flaps down, each engine needed up to 12,000 lbs/hr of fuel flow to stay on speed and on the correct profile. As speed, configuration, and wind change, thrust needs to be adjusted; because the autothrottles were controlling thrust, they made four corrections in fairly rapid succession, the last occurring 720 feet above the ground. Though commanded higher, the right engine initially rolled back to nearly idle; the left engine rolled back just 7 seconds later. The pilots first became aware of the problem at about 590 feet above the ground with 48 seconds until touchdown, when they noted the throttles were no longer parallel with each other but split apart as the right engine started to roll back. At that point the crew was trying to understand the situation, analyze it, and take the safest course of action. Given the time constraints, they did well, obeying the aviation dictum to aviate, navigate, and communicate, in that order. The most important task facing any crew is to fly the plane. It seems obvious, but numerous aircraft have crashed because the entire crew was so focused on solving a problem that no one was flying the plane (the most famous example being Eastern Flight 401, an L-1011 that crashed in the Everglades on December 29, 1972). In this case, the crew successfully flew the plane while analyzing the dual engine rollback, a scenario they had not been trained for because it was deemed so unlikely.

As the crew analyzed the situation, it became evident that the aircraft would not reach the runway. The Captain recognized that retracting the flaps from the normal landing setting of 30° to 25° would extend the glide closer to the runway and help avoid obstacles in the path of the aircraft. The net effect of the reduction to 25° was to lengthen the glide by about 164 feet, though both scenarios would have left the aircraft within the airfield boundary; the extra distance helped the aircraft fly over lighting installations that could have done much greater damage to the plane. The trade-off the Captain faced with this split-second decision was that while raising the flaps to 25° lowered drag and allowed the aircraft to glide farther, it also lowered lift by reducing the effective area and camber of the wing. This in turn raised the stall speed of the aircraft. An aerodynamic stall is technically defined as exceeding the critical angle of attack, but for the purpose of this discussion of airliners that are making gentle maneuvers, the stall speed of the aircraft is the speed below which there is not enough lift produced by the forward motion of the wing through the air to support the weight of the aircraft. When an aircraft attempts to fly below this speed, the plane will become much more difficult to control (and can be somewhat unpredictable) and will start to descend rapidly. Avoiding a stall was of paramount importance to the crew of Flight 38. For the aircraft weight at the time of the accident, the Flaps 30° stall speed was 104 knots while the Flaps 25° stall speed was 106 knots. In either case, a warning device called a stick shaker operates at 4 knots above either stall speed and vibrates the control yoke the pilot is holding to fly the plane to give a tactile and audible warning that speed is dangerously low. About 200 feet above the ground with 10 seconds to touchdown, the stick shaker activated, forcing the pilot to lower the nose of the aircraft to prevent a stall. The crew managed the available energy (kinetic and potential energy of height) as well as possible in the brief time available to take action, and had avoided the major obstacles in the path of the aircraft, heeding the famous advice of test pilot Bob Hoover: "If you're faced with a forced landing, fly the thing as far into the crash as possible." The crew clearly managed to aviate and navigate well; communicating was a distant third priority, and the crew had only enough time to transmit a Mayday call to the control tower 3 seconds before impact. There was no time to warn the passengers or Flight Attendants. The impact was very hard at 25 ft/sec, or about 1,500 feet per minute. A normal, comfortable touchdown would be around 100-200 feet per minute by way of comparison; although hard landings have varying definitions from aircraft to aircraft and between some regulatory regimes, a commonly used definition of a hard landing is 360 feet per minute or greater. Above that value, inspections are required before the plane can fly again. In this case, the landing was hard enough to result in landing gear collapse, and the aircraft could not be economically repaired and was scrapped. This crew was faced with a serious situation and had little time to react; they made the best of a bad situation thanks to good airmanship and excellent training.

So what caused the engine rollbacks in the first place? Recall that British Airways Flight 38 had low fuel flows throughout most of the flight and was operating in extremely cold temperatures. All aviation fuel has water suspended in it, and the fuel on board Flight 38 met all specifications. Operating in these extreme cold temperatures for an extended time allowed ice to accumulate on the pipes and components of the fuel system. That is undesirable but not unusual, which is why jet aircraft have methods of heating the fuel prior to delivery to the engines. Some aircraft use hot air tapped off of the engines to heat this fuel as necessary, normally in combination with a fuel/oil heat exchanger (FOHE). The FOHE is an elegant bit of design because hot engine oil needs to be cooled and cold fuel needs to be heated. Most newer engines, including the engines on the B-777, depend solely on an FOHE that is computer regulated to control fuel and oil temperature within safe parameters. Unfortunately, it was not realized during the design stage that if fuel flow was steady and low through most of the flight, ice could accumulate on one side of the FOHE and could not be heated rapidly enough to melt it without causing a thrust restriction. It was found that reducing the engine to idle power for 30 seconds would clear the occlusion (this was obviously not practical during Flight 38's time line, nor did the pilots have any knowledge of this phenomenon). Reducing thrust to idle allowed the stasis in the fuel system components to be broken and allowed the hot oil to dwell in the FOHE for longer melting the ice, as it was designed. Once this was understood, operators of B-777 aircraft with this type of engine began operational procedures during the cruise and descent phases of flight to cycle the thrust of the engine to ensure that this accident scenario would not recur. In the long run, worldwide regulatory agencies required a redesign of the FOHE to physically prevent a recurrence of such an accident.

Landing

Statistically, 24% of commercial jet accidents with fatalities occur during approach and 24% occur during landing. As we saw in Chapter 8, even if destroyed, a modern commercial jet still has great potential to protect the people. Asiana Airlines Flight 214, a Boeing 777, was destroyed by impact and fire while landing on July 6, 2013, in San Francisco. Three of the 307 passengers and crew were fatally injured; two of them, however, would have survived if wearing seatbelts.

Unique to this flight, a trainee pilot was making his first Boeing 777 visual approach and landing. The pilot was training to become a Boeing 777 Captain. The trainee had 9,684 total flight hours, including well over 3,000 hours as a Captain (mostly in the Airbus A320), and had completed additional 777 training as an observer and in a simulator. The instructor pilot in the right seat had 12,307 flight hours, including 3,208 as a 777 Captain. Coincidentally, the instructor pilot was on his first flight as an instructor. A second relief crew was present for the 10.5-hour flight from Korea.

Obviously, a plane must slow and descend to land. The most efficient descent profile occurs with the throttles at idle. In the real world, most descents are made in a series of steps to lower altitudes and slower speeds in accordance with instructions from air traffic control. Additional speed is lost (increasing descent rate and drag) when flaps, speedbrakes, and landing gear are extended. Approaching San Francisco International Airport, Flight 214 was cleared by controllers to slow to 210 knots. Between 9,000 and 6,000 feet, the speedbrakes were extended for 54 seconds. Speedbrakes extended on the 777 at idle thrust and 210 knots will increase descent rate from 1,000 feet per minute to about 2,300 feet per minute. High descent (sink) rates are not safe close to the ground; most airliners descend at about 700-800 feet per minute on final approach. For that reason, speedbrakes are not allowed below 1,000 feet; likewise, for passenger comfort (to reduce turbulence and noise), speedbrakes are not recommended in conjunction with flap settings above 5° on the 777.

Sink rates are reduced as the plane gets closer to the ground. Standard practice: less than 3,300 feet per minute between 5,000 and 2,000 feet, and less than 1,500 feet per minute between 2,000 and 1,000 feet. Below 1,000 feet, the plane must meet additional rules defining a stabilized approach or immediately begin a go-around, which is essentially a landing do-over.

At 4 minutes and 33 seconds before impact, Flight 214 was 14.1 miles out, 4,800 feet high, with airspeed 215 knots, and descending at 900 feet per minute; the plane was cleared to reduce airspeed to 180 knots until 5 miles out. The flaps had just been extended to 1°. The 777 has six flap settings with maximum speed limits, as shown: 1° (255 knots), 5° (235 knots), 15° (215 knots), 20° (195 knots), 25° (185 knots), and 30° (170 knots). At excessive speeds, windblast will damage the flaps; the minimum speeds are defined by the approach to stall angle of attack for each configuration. The maximum and minimum speeds permissible for each flap configuration are displayed on each pilot's primary flight display (see Figure 5.1). The flaps were extended to 5° immediately before the landing gear was lowered; the plane was then about 8.5 miles out, 188 knots, 3,500 feet, and descending at about 1,000 feet per minute. The landing gear was extended (must be less than 270 knots) a bit early because the plane was about 1,000 feet above the 3° glideslope, and the landing gear adds a large amount of drag to the airplane, enabling the pilots to descend rapidly to a normal approach profile. Later, Flaps 20 were extended at around 1,850 feet, quickly followed by Flaps 30 at 1,330 feet just 22 seconds later.

Construction that moved the runway threshold farther from the seawall had also temporarily disrupted the instrument landing system (ILS) glideslope antenna. This electronic glideslope defines the vertical component of the ILS system; an inoperative glideslope requires the pilots to revert to a nonprecision approach or a visual approach; in this case, because the weather was good, the flight conducted a visual approach. Onboard computers also derive a vertical path that approximates the ILS glideslope; this is displayed in the cockpit by the vertical path indicator. The precision approach path indicator (PAPI) system—a system of four lights next to the runway—also gives pilots a visual indication of their relationship to the correct glide path. PAPI lights are visible in clear weather during the day from at least 5 miles out, and 20 miles at night. Pilots are also expected to maintain a mental picture of the glideslope, backed up by a basic mathematical formula generically called "the 3:1 rule," in which they fly to have 300 feet of altitude for every mile out (e.g., 2 miles from the airport, the altitude should be 600 feet above the ground).

Flight 214 was consistently about 500 to 1,000 feet high during approach; this aspect of the accident was later studied by simulation. Ten landings with a proper approach—defined as 1,650 feet altitude, airspeed 180 knots, and sink rate of 850 feet per minute at 5 nautical miles from the end of the runway—were compared to the accident profile of 2,100 feet, 175 knots, and 1,000 feet per minute sink rate. The stabilized approach criteria were easily met in the simulator when starting with the standard profile. For the accident profile, four simulated flights failed to stabilize the approach (and had to fly a go-around); the remaining six flights had difficulty stabilizing the sink rate.

Each company defines a stabilized approach in their manuals, and most such descriptions are similar. At Asiana, instrument approaches must meet the stabilized approach criteria at 1,000 feet. Flight 214, flying under visual flight rules, had to meet the criteria at 500 feet above the ground. The procedure requires the Pilot Monitoring to call out "500"; the Pilot Flying must respond "stabilized" or "go-around." (Allowing for an incapacitated pilot, the procedure mandates that the Pilot Monitoring take control if the Pilot Flying does not respond to a second callout.) The "500" callout did not take place, and Flight 214 never met the stabilized approach criteria, which include: sink rate less than 1,000 feet per minute, airspeed within +10 knots or –5 knots of the target airspeed, on the correct glideslope, only minor adjustments to pitch and roll are required, and—the most important requirement for Flight 214—that throttle setting "is appropriate for the airplane configuration," with the engines continuously "spooled up" (i.e., not at or near idle). Flight 214's Pilot Flying (the trainee Captain) shut off the autopilot at around 1,600 feet (the overwhelming majority of flights land without the autopilot, a phenomenon Captain Hedges explains in Chapter 8). Through a fast-paced series of errors and misunderstandings related to the complexity of the aircraft's automated systems, the aircraft was put in a high-drag, low-thrust state; none of the pilots understood that they were in a mode that would not hold the approach airspeed using the autothrottle, nor did they notice the rapidly decreasing airspeed until it was too late. At 500 feet, the plane was still high but was sinking fast; by 300 feet, the aircraft was going below the correct glide path and getting dangerously slow. Because of decreasing airspeed (and increasing sink rate), the pilot tried to compensate by rapidly pitching the nose up—another violation of the stabilized approach criteria.

Target speed for the approach was 137 knots. This speed is V_{ref} + 5 knots. V_{ref} is 1.23 times the plane's stall speed, which depends on the plane's weight and flap setting. For 423,360 lbs and Flaps 30, the stall speed is estimated to be 107 knots. (The weight is estimated from the plane's performance.) On a "perfect" landing, V_{ref} occurs as the plane crosses the runway's threshold at an altitude of 50 feet.

Table 9.1 picks up the action at 1,000 feet. The runway light system, or precision approach path indicator (PAPI), displays two red lights and two white if the plane is on the correct glide path. The plane is below the proper glide path with more than two red (R) lights, and above the proper glide path with more than two white (W) lights; the correct PAPI configuration is WWRR. Note the rapid PAPI and speed changes in Table 9.1. Procedures require the observer pilot (part of the relief crew), who on this flight was sitting in the jump seat, to also monitor the approach. The statement "Sink rate sir" (51.7 and 45.2 seconds before impact, as the plane descends from 900 to 700 feet) indicates that the observer pilot is concerned about excessive sink rate. The observer pilot callouts coincided with the maximum recorded sink rate of 1,776 feet per minute—a value that greatly exceeds the 1,000-foot-per-minute limit in Asiana's flight manual for descent below 1,000 feet.

Asiana's policy calls for an immediate go-around if four red PAPI lights are visible, as occurred at least 20 seconds before impact at an altitude of 220 feet. This is a clear indication that the plane is too low for a stable approach. The instructor pilot initiated a go-around about 7 seconds before

Table 9.1. Flight 214 Data below 1,000 Feet

Time to Impact (seconds)	Altitude (feet)	Airspeed (knots)	Details
55.3	1,000	150.5	PAPI displaying WWWW
51.7	917	146.7	Observer Pilot says, "sink rate sir"
51.2	904	147.2	Pilot Flying says, "yes sir"
45.2	723	146.0	Observer Pilot says, "sink rate sir"
35.4	500	136.7	PAPI changes to WWWR
30.5	404	134.0	Pilot Monitoring says "on glide path sir"; PAPI changes to WWRR
26.2	331	129.7	PAPI changes to WRRR
19.3	219	121.9	PAPI changes to RRRR
16.7	180	117.6	Pilot Monitoring says, "it's low"
15.5	165	119.2	Pilot Flying says, "yeah"
7.5	90	109.8	Pilot Monitoring says, "speed"
7.1	86	109.1	Right engine thrust lever advanced
6.6	81	108.4	Left engine thrust lever advanced
3.9	46	103.6	Stick shaker engaged
2.5	29	103.8	Pilot Monitoring says, "go around"
1.7	21	104.7	Stick shaker disengaged
0	5	105.5	Impact

Note: Correct PAPI configuration is white-white-red-red (WWRR).

impact. Studies by the National Transportation and Safety Board and Boeing concluded that the go-around could have been completed if initiated at least 11–12 seconds before impact. Because of the turbomachinery's rotational inertia, there is a delay between advancing the throttle and increased speed. A 4-second delay is seen in Table 9.1 between increased throttles and airspeed finally ticking up just 0.2 knots; this illustrates why it is important for jet aircraft to have the engines spooled up near the ground.

N1 describes the rotation of the turbofans and directly correlates with power. N1, displayed in the cockpit as an indicator of engine performance, is usually expressed as a percentage of full power. About seven seconds before impact, the recorded N1 for Flight 214 during approach with throttles set at idle was 24%. About four seconds after the throttles were advanced, N1 was approximately 50% on each engine. About three seconds later (during impact), N1 was 92%; the plane's speed had increased from its low of 103.6 knots to a speed at impact of 105.5 knots. The flight manual describes an example N1 of 57.9% during approach with Flaps 30, landing gear down, 142 knots, and landing weight of 450,000 lbs.

Surprisingly, the plane did not stall. The stick shaker activated, indicating an impending stall at 103.6 knots, and deactivated at 104.7 knots. The stick shaker algorithm depends on angle of attack (AOA); Mach number; airspeed; slat, flap, speedbrake and landing gear position; and engine thrust. The discrepancy between predicted stall at 107 knots and no stall at 103 knots is explained by pilot inputs. The pilot was trying to compensate for reduced speed by increasing the AOA. The expected airplane pitch at final approach for a 450,000 lb 777 is about 0°. The AOA is about 2° greater than the plane's pitch. Flight 214's AOA gradually increased to about 15° four seconds before impact. (At impact, the control column was at the full aft stop, with both pilots exerting over 100 lbs of column force.) Although the plane is safe to fly at a much higher AOA, the simulations are calibrated with test flights. Boeing does not consider it safe to fly a 400,000 lb plane close to the ground with AOAs of less than –2° or greater than 10°. Because lift owing to ground effect increases closer to the ground (see the discussion of ground effect in Chapter 6), the simulation and stall prediction is not exactly correct for AOAs greater than 10° close to the ground.

An idealized approach is shown along with the impact orientation in Figure 9.1. The landing gear hit the seawall first and cleanly broke off, as designed, 600 feet short of the runway's threshold. The tail section struck next and broke off at the aft pressure bulkhead. (The three fatalities were seated in the last two rows and close to the point of impact.) The airplane slid along the runway, became partially airborne, rotated about 330°, broke off both engines, and finally came to rest about 2,400 feet from the seawall. The left engine tumbled about 600 feet north of the mostly east-west runway. The right engine came to rest against the fuselage and ignited a fire. The landing gear and engines are connected with mechanical fuse pins designed to break away cleanly and protect the fuel tanks; both functioned well, and the overall impact resistance of the aircraft was praiseworthy.

Beginning 12 seconds after impact, the wreckage was enveloped by a dust/smoke cloud for about 26 seconds, making it difficult to determine exactly when the plane came to a stop. It's believed the fuselage slid to a stop about 2,400 feet past the seawall in about 16 seconds.

During impact, six people were ejected from the plane. This included four seriously injured Flight Attendants sitting in the back of the plane and two fatally injured passengers not wearing seatbelts.

It was not a perfect evacuation. Most likely, the 26-second dust cloud confused things. Certainly, having only 4 of 12 Flight Attendants available to help (the rest having been ejected out the back or injured) added to the confusion, as did the two evacuation slides that opened inside the aircraft,

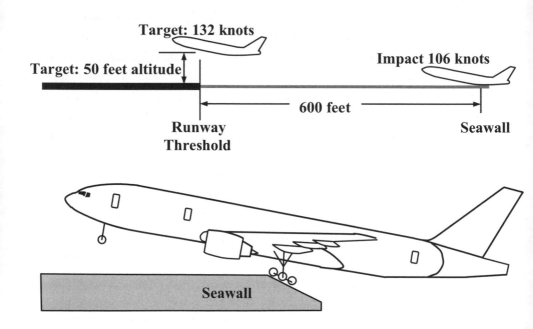

Figure 9.1. Idealized approach is shown compared to Flight 214's orientation on impact.

Approach and $f = ma$

We can estimate time and distance traveled during approach with $f = ma$. As explained in Chapter 7, approach is mostly an unpowered glide. Without engine thrust, the only forces acting on the plane are gravity and drag.

Gravity tries to accelerate the plane in its direction of motion. Assuming a 3° glideslope, the acceleration is the same as if the plane were sliding (without friction) down a 3° incline. Working out the trigonometric equations gives a simple result: the gravity force equals a percentage of the weight equal to the value of the glideslope angle; in this case, the gravity force equals 3% of the weight of the plane. Assuming a 450,000 lb weight, the gravity force is 0.03 × 450,000 lbs = 13,500 lbs of force accelerating the plane along the direction of motion.

If the plane is flying at its optimum orientation, the drag force can be estimated by the lift-to-drag ratio. A typical ratio for large commercial jets is 17. Because lift must equal weight on straight and level flight, a meaningful estimate of the drag force is 450,000 ÷ 17 = 26,500 lbs of force decelerating the plane. The net force trying to slow the plane is 26,500 – 13,500 = 13,000 lbs of force trying to slow the plane as it descends, as shown in the figure.

We will write $f = ma$ in the direction of flight, and therefore we are only concerned with the forces acting on the airplane in that direction. Forces perpendicular to the direction of flight such as lift are ignored, which is by definition a force perpendicular to the direction of flight.

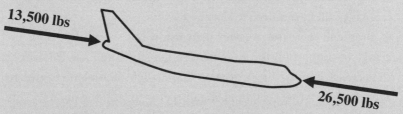

13,500 lbs

26,500 lbs

Estimate of gravity and drag forces acting on the plane during approach.

Applying a 13,000 lb decelerating force to our 450,000 lb plane, the deceleration is according to $f = ma$ (don't forget to convert weight to mass by dividing by the acceleration of gravity, or 32 ft/sec^2).

$$13,000 = (450,000 \div 32) \times \text{acceleration}$$

Acceleration = 0.924 ft/sec^2, or about 2.88% of normal gravity, a gentle deceleration (in the direction of flight) compared to the proverbial falling apple.

If we double the weight of the plane, the mass, drag, and gravity forces all double and we get the same result. We therefore conclude that the results are independent of the plane's weight. Captain Hedges confirms that for descent planning in the real world, pilots should use the 3:1 rule wherein the plane is expected to lose 1,000 feet of altitude every 3 miles, independent of aircraft weight.

Starting at the maximum permitted speed of 250 knots below 10,000 feet, the plane must slow to approximately 150 knots for touchdown. Let's assume the plane is slowing from about 250 knots to 150 knots, or from about 420 ft/sec to around 250 ft/sec. The plane must reduce its speed 420 − 250 = 170 ft/sec.

Assuming a constant deceleration of 0.924 ft/sec^2, a slowing of 170 ft/sec will occur in

$$170 \text{ ft/sec} = 0.924 \text{ ft/sec}^2 \times 184 \text{ seconds}.$$

If the plane slows from 420 ft/sec to 250 ft/sec, its average speed is 335 ft/sec. The plane will travel 335 ft/sec × 184 seconds = 61,600 feet, or about 10 nautical miles. This estimate agrees well with the pilot's rule of thumb: 1 nautical mile of distance traveled is needed to slow 10 knots, and therefore approximately 10 nautical miles horizontally will be required to slow 100 knots.

Drag will increase in lower denser air, but so will lift. Both increase in proportion to the increased density. The nose is lowered to decrease both (if lift is greater than weight, the plane begins to accelerate up—the exact opposite of trying to land the plane!). This suggests that the glide ratio is mostly constant for all altitudes. (Captain Hedges notes: Although this is generally true at a constant airspeed, in practice, the glide ratio is sensitive to airspeed, and there is an optimum speed for every weight and aircraft configuration.)

Drag (and lift) will increase as the flaps are progressively extended. The actual distance attainable in a glide is something less than the calculated value.

blocking aisles. The evacuation did not start for 1 minute and 33 seconds. Evacuation slides opening inside the cabin is rare, but not unprecedented, most recently occurring on a United Airlines Boeing 737 during flight. Everything on a plane is designed for Federal Aviation Administration–certified crash loads. Unfortunately, it is always possible to exceed the design crash loads, and that was the case with Asiana 214. Interestingly, no defects were found in the slides that opened inside.

One ejected passenger, possibly already dead, was run over by a fire truck. Five passengers were too injured to evacuate; a Flight Attendant went to help but was forced back by fire. Firefighters with breathing equipment rescued the injured ahead of the fire. The burned-out wreckage of Flight 214 is shown in Figure 9.2.

If there is only one lesson from the Asiana 214 accident, it's that large aircraft have considerable momentum, and their direction and speed are not easily changed and must be properly managed. During approach, the pilot monitors a slew of constantly changing variables, including airspeed, sink rate, and altitude; the target values are customized for each runway and constantly change with distance from touchdown. For example, the pilot can lower the nose or add engine thrust to increase speed. Angle of attack and flap settings interact with all of the above. It's not only a lengthy list of variables, but complex interactions exist between them. Conditions on the ground change, including: visibility, winds (head-, tail-, or cross-), ambient air density, and in the case of Asiana Flight 214 airport equipment operating (or not operating). Flying a specified glide path with abruptly changing condition has made approach the most challenging phase of flight. Historically, controlled flight into terrain (CFIT), mostly associated with limited

Figure 9.2. The wreckage of Flight 214. Wikimedia/National Transportation Safety Board

visibility and pilot distraction, has been an enormous problem that has not been completely eliminated. The modern (and considerably safer) solution is to manage the problem with software or, as explained by Captain Hedges, sometimes multiple software selections in the form of various automation modes.

Every approach is different, and the variabilities inherent in these different procedures require the flexibility of multiple automation modes. Understand that the pilot is still flying the plane but is using the various software modes as tools to safely and precisely operate the aircraft. None of this occurs in a vacuum, and at busy commercial airports, pilots can expect instructions from air traffic control (ATC) several times a minute. ATC is responsible for the safe separation of aircraft and for sequencing them for maximum efficiency. As such, ATC will give clearances, some of which are frequently quite complex. Controllers are good at keeping traffic moving and will assign speeds, altitudes, headings, and other clearances (e.g., "fly heading 300, maintain 2,800 feet until established on the Localizer, cleared the ILS 27R, maintain 180 knots until Depot, contact tower at Depot on 123.85") in a fast-paced aerial ballet. In the cockpit, the pilots not only have to hear and understand everything the controller said, but also must read it back to the controller to make sure the instructions were correctly understood, all while flying the airplane and complying with the clearance. The controller can change the clearance at any time, most commonly with different headings or speeds, but everything including the landing runway and kind of approach can be rapidly altered. To meet these challenges, pilots need capable machines to help them manage their workload. The descent and approach phases may require multiple modes depending on the situation: a clearance to expedite a descent and then slow to a specified speed will be handled with a different combination of mode and configuration (flaps, landing gear, etc.) changes than a clearance to slow first and then descend, while the final approach segment will be handled differently still; these automation modes and the software behind them are key to keeping the system operating reliably and safely in all kinds of weather. The configuration of the aircraft must be carefully managed by the crew, and at different speeds and configurations the aircraft has different handling characteristics. Just because they are executing an approach doesn't mean the pilots can stop analyzing other things like winds, weather, or changing runway conditions, and on every approach pilots must have a go-around plan in case a safe landing can't be assured (or if ATC directs it).

Some may think that complete system automation is the answer, and one day it may be. For the time being, it's impossible because there are no suitable software interfaces that can get the instructions from ATC into the plane in a timely enough manner. There are examples of various types of drones now flying, some of which are autonomous, and perhaps in the longer term this will be a valid approach, but for now and the foreseeable future, neither the hardware nor software is adequate to fly passengers into busy airports with the reliability certification that authorities require. In commercial aviation, it's not acceptable for the system to work 99.9999% of the time, because reliability is literally a matter of life and death. (A cell-phone network, with its frequent dropped calls, illustrates the problem and is nowhere near reliable enough.) As a practical aside, no company would currently insure an airline that wanted to put passengers into a pilotless aircraft. If there are any medium-term applications for reducing the pilot in the control loop, it's more likely in the cargo realm, though the software isn't there yet and isn't flexible or reliable enough.

Captain Hedges Explains: Automation and Mode Awareness

As aircraft became more automated, aircraft capabilities increased dramatically. On modern designs like the 777, the autoflight system is extremely reliable and capable. On a suitably equipped approach, such aircraft are fully capable of landing themselves without the pilots required to see the runway prior to touchdown. They are also capable of navigating to any point in the world with great accuracy and can conduct vertical profiles to cross certain points at a specified altitude and speed. These advances are wonderful and reduce pilot workload, but a corollary is that pilots must have greater insight into how the aircraft is achieving its goals and what modes it is operating in at all times. Aircraft manufacturers and airlines speak of "levels of automation," and while all planes are different and airlines use varying terminology, the training that pilots receive for these aircraft emphasizes the various levels of automation and when and how to use them. Many people think of the autopilot as an all-or-nothing device, that it's either on or it's off. In older aircraft that was true, but newer aircraft have multiple different levels of capability.

Pilots still get trained on "stick and rudder" skills, or flying the aircraft by hand with the automation turned off. Manual flying skills will always be important, and all airliners can and are flown manually by the pilot with reference to basic instruments and the ground. Pilots also get training with the automation working to the maximum degree with programmed lateral and vertical paths and automatic landings in extremely low visibility. Modern airliners have capable automatic systems, though they all come with restrictions on their use, and understanding the operation and limitations of these systems is a huge emphasis in airline operations.

Contrary to popular belief, between fully manual (no automation) "hand flying," where the pilot makes all control inputs in a way similar to the previous century's pilots, and fully automatic operations, where the aircraft optimizes its lateral and vertical profiles on the basis of various inputs and can conduct an automatic instrument landing system (ILS) approach and landing without the pilot touching the controls, there are various levels of automation available to pilots. Matching the appropriate level of automation to a given situation is a challenge to learn and a challenge to teach, and avoiding automation errors has become a huge emphasis in airline training programs. Knowing when to use a given feature or to manually intervene is a learned skill, and it's one reason even experienced pilots (like the Asiana 214 trainee Captain) fly with an instructor for a few trips when they transition to a new plane. The simulator is a great and realistic training device, but nothing compares to real-world operational experience. Examples of when it's appropriate to use different levels of automation are most obvious at the extremes of operation. If the aircraft and airport are suitably equipped (and pilots are certified to do so), planes can execute an ILS approach to an automatic landing in zero visibility: the pilots never need to see the runway before landing (in reality, while we don't have to see the runway to land in these situations, in practice there is a minimum of around 300 feet of visibility required because we need to see to taxi to the gate; if you're in a B-747, that's just over one plane length). This is obviously a time when full automation is required: if a pilot can't see the runway before touchdown, it would be impossible to safely land the aircraft manually.

On the other extreme, if an aircraft will be landing in a significant crosswind, a manual landing will be required: autopilots have a maximum crosswind limit for automatic landings that may be as low as 10 knots, whereas the aircraft may be able to be manually landed with more than 30 knots of crosswind. Between these two automation extremes are a variety of capabilities and modes available to pilots; it may be helpful to think of this as a "third way" of flying the aircraft, with some features fully automated and others being done manually. The variations are nearly endless, and different pilots may handle situations with different modes or techniques. This is perfectly acceptable so long as the crew complies with procedures (e.g., stabilized approach criteria). It is possible, for instance, to turn off the autopilot and flight directors (explained on page 269) and fly a visual approach by looking out the window at the runway. In that situation, the pilot has the option to use the autothrottles to maintain a selected speed or to disconnect the autothrottles and maintain airspeed by manually manipulating the throttles (in either case, the pilot must monitor speed and sink rate). Regardless of the automation level or mode used, it's important for pilots to understand what the aircraft is doing and why, and to be ready to disconnect the automation and revert to manual flying skills with no notice.

Flight Mode Annunciator

As airliners became more sophisticated, it was important to have a single place on the pilot's instrument panel to see and understand the automation status of the aircraft. This concept has evolved into a display called the flight mode annunciator (FMA), and pilot monitoring and understanding of the modes on the FMA are central to safe flight and are emphasized in training programs. In the 777, like all current Boeing aircraft, the FMA is at the top of the primary flight display (PFD), immediately above the attitude director indicator (ADI). At some airlines, FMA monitoring is deemed so critical that every change to it is verbalized by a crewmember. The following discussion is generalized and does not attempt to educate readers on every mode or submode available or certain advanced nuances of any particular autoflight system. Each aircraft is slightly different, and lessons applicable to one aircraft manufacturer or type aren't necessarily germane to another.

In the 777, similarly to other advanced Boeing aircraft, there are three primary methods of vertically maneuvering the aircraft: vertical speed (V/S), flight level change (FLCH) (explained on page 267) and vertical navigation (VNAV). If the autopilot is engaged, it will automatically follow the commands of whatever mode is active; if the autopilot is off, the pilot must manually maneuver the aircraft to comply with the desired flightpath. In any given flight, pilots may frequently change between modes as objectives change. VNAV is the highest level of automation available and complies with complex altitude and speed profiles, allowing pilots to program the system to be at a certain altitude and speed at a given point in the flight plan and then change speed and altitude to comply with subsequent restrictions. VNAV was not a factor in this accident. V/S was previously discussed in the British Airways 38 Boeing 777 accident, and allows the aircraft to climb or descend at a specified rate while maintaining a specified airspeed. This is a useful and commonly used mode, and initially this was the mode Asiana 214 was operating in during the approach. FLCH is another common mode and varies in its functionality depending on whether the altitude preselected by the crew is above or below the crew; this is key to the accident and is discussed in the summary below.

Why Are Multiple Automation Modes Needed?

It's critical that pilots are proficient in all automation modes because a single arrival may necessitate using several different modes (as the crew of Asiana 214 had to do), all of which have different characteristics. As an example, from the top of descent to somewhere in the terminal area, pilots would typically use VNAV, as air traffic control (ATC) will likely mandate crossing certain points at specified altitudes and speeds, and this mode (or its Airbus equivalent, "Managed Descent") will ensure those restrictions are made. If a large altitude change is required and no crossing restrictions are specified (especially if there is not much distance to get down), pilots will want to make sure the engines are in idle and would likely pick FLCH (the Airbus equivalent is "Open Descent"), which will descend the aircraft in idle by default. Remember that an airliner is optimized for efficient flight, minimizing drag and maximizing lift on the least amount of fuel possible; the corollary of that is that airliners are challenging to descend and are extremely difficult to descend and slow at the same time. Every aircraft is different: the B-757 and A321, for instance, are aerodynamically clean and challenging to descend; the MD-80 is older and has high allowable flap and slat extension speeds and effective speedbrakes, making it much easier to descend over a given distance. If there is a small altitude change required (generally less than 1,000 feet; see Chapter 8), pilots will frequently choose V/S because it is smoother and more comfortable for passengers. It's also a required mode at some airlines on some nonprecision (i.e., non–instrument landing system, or ILS) approaches. The takeaway here is that as a plane descends, numerous different modes may be used: at first, VNAV will most likely be used to ensure certain crossing restriction clearances are complied with, after which ATC may step the aircraft down in large or small altitude increments, where FLCH or V/S would be used, respectively. ATC may also tell an aircraft to expedite its descent, when FLCH would be the norm, probably with speedbrakes, and occasionally will specify a rate of descent as well, when V/S would be the obvious choice (that's what it's there for). Every arrival is different, and this real-world flying is why instructors are so important to have along when pilots transition to a new aircraft type, as each plane has different characteristics. Ultimately, if there's an ILS in use, pilots will end up flying the electronic glideslope down final approach. Depending on aircraft capabilities and features for non-ILS approaches, the likely vertical modes would be V/S or VNAV; it should never be FLCH (or "Open Descent" in the Airbus), as those modes allow the thrust to be at idle, which is not permitted by stabilized approach criteria. If a plane is cleared for a visual approach (as Asiana 214 was), the other option is to turn off automation (to the degree airline-specific procedures allow), including autopilot, flight directors, and autothrottles ("autothrust" in Airbus aircraft), and fly the way pilots have for more than a century, mentally calculating a 3:1 descent, controlling speed with power and aircraft configuration (flaps, landing gear, etc.), and looking out the window.

As Asiana 214 descended through approximately 2,300 feet, the pilots set 3,000 feet into the mode control panel (MCP), which the crew uses to interface with important cockpit automation, including autopilot, autothrottles, and flight directors. Flight directors are bars superimposed on the attitude director indicator that generate commands in pitch and roll to direct the pilot how to maneuver the aircraft; if the autopilot is on, it will follow the same commands displayed by the flight directors. With the autopilot off, the pilot must fly the aircraft manually either with or without the flight directors. Setting 3,000 feet in the MCP was done in case the aircraft had to do a go-around: it would be the procedural altitude the plane climbed to in that event. When this selection was made, the plane was about 500 feet above the desired glide path, descending in the V/S mode with the autopilot on, and was attempting to slow to final approach speed. In an effort to descend more rapidly back to the proper glide path, one of the pilots (probably the trainee Captain) pushed the FLCH button on the MCP. FLCH does two primary things (the logic has been simplified a bit for this discussion): if the altitude in the MCP window is below the aircraft, the autothrottles will retard and the aircraft will descend at the selected speed; if the altitude in the window is above the aircraft, the autothrottles will advance to climb power and the aircraft will climb to the altitude in the MCP window. The throttles can be overridden by the pilot and will then enter a "Hold" mode, in which their servo motors are de-energized and the throttles will not move without pilot intervention. The pilots expected FLCH to retard the throttles and to make the aircraft descend faster, but they had forgotten that the MCP altitude was now set *above* their current altitude. The autothrottles dutifully advanced to climb thrust (a high power setting), and the plane pitched up to climb to 3,000 feet. The trainee Captain had good instincts, as he quickly disconnected the autopilot, manually retarded the throttles back to idle, and lowered the nose.

What nobody noticed is that in the process of manually retarding the throttles to idle, the autothrottles had entered the Hold mode (as designed), in which the pilot has sole responsibility for moving them to adjust the speed of the aircraft. This isn't as much of a problem if the autopilot is left engaged, as the aircraft will simply use pitch to maintain the proper airspeed complying with the aircraft generated commands. When the autopilot is disconnected, however, the throttles remain in Hold mode until the aircraft captures the altitude in the window of the MCP, and the power will remain wherever the throttles are placed by the pilot (in this case, idle) until that altitude is attained. If a pilot pitches the aircraft up with the throttles in Hold mode at idle, the plane will begin to slow. This is a complex automation interaction that was not well understood by Asiana pilots, the accident investigation revealed. The stage for the accident was now set: the aircraft was fully configured for landing (a high-drag state), the descent rate was high, and the throttles were in idle, yet nobody in the cockpit was aware that the autothrottle system was not protecting their speed. The trainee pilot called for the flight directors to be turned off and the aircraft crossed through the desired glide path and began to get low, still in idle. Ironically, having both flight directors on or off is an emphasis item in Airbus aircraft (the trainee Captain most recently flew the Airbus A320), and in this case the trainee Captain turned his flight director off (a wise choice because it wasn't providing useful information), but the instructor did not. One way in the 777 to force the autothrottles to wake up and enter the Speed mode, in which they will maintain the commanded speed (this is the mode the Asiana pilots thought they were in), is to turn off all autopilots and flight directors: this is why Airbus emphasizes that both pilots either use the flight directors and obey them, or turn them both off (it was common practice at Asiana in the 777 fleet to turn off both flight directors during a visual approach, and then have the Pilot Monitoring, in this case the instructor pilot, turn his back on). If the instructor had turned off his flight director, the autothrottles would have come out of Hold mode and engaged to maintain the selected speed (in this case, 137 knots), the desired objective. Instead, the crew never grasped that they didn't have speed protection and trended lower and slower at the end of this unstable approach. A clear go-around call earlier in the sequence would have been in order, and one result of the accident is an increased emphasis on stabilized approaches and timely go-around decisions. Asiana 214 was a tragic but preventable accident, but the system is safe overall: there have been only two accidents in over 16 million 777 landings, and at least two of the three fatal injuries were preventable by proper seatbelt use. Flight mode annunciator awareness has been greatly emphasized in high-technology cockpits since Asiana 214, and airlines have done more to educate 777 (as well as 757, 767, and 787) crews of this potential automation trap, making a similar accident less like to recur.

Two Ways to Ruin a Landing

Asiana 214 is an example of an "undershoot," or accident in which the aircraft hits the ground short of the runway, normally after getting low and slow on the approach. These accidents tend to be relatively high-energy accidents because the aircraft are traveling at substantial speed at the beginning of the impact sequence. Another common type of landing accident is an "overshoot" or "overrun" in which, after landing, an aircraft departs the far end of the runway. These accidents are generically called "runway excursions"; most of these accidents involve departing the end of the runway, although another type of runway excursion occurs when a plane departs the side of the runway. This is relatively uncommon but most often involves slippery runways in combination with a crosswind.

Landing overruns have a lot in common with overruns following a rejected takeoff (see Chapter 2): the impact is normally lower energy than an undershoot as the plane has decelerated as much as possible on the runway and then leaves the end of the runway. These accidents are normally survivable, although the aircraft may be destroyed. Where undershoot accidents normally are spawned by a low and slow approach, overshoot accident accidents are frequently the result of a high and fast approach or a slippery runway surface. When a high and fast approach is combined with a slippery runway, the recipe for an overrun is complete, as was the case for Air France Flight 358 on August 2, 2005.

Air France Flight 358

About one-third of all large jet accidents (and the largest single accident category) are landing overruns, where the plane runs off the runway during landing. These accident scenarios are similar to the rejected takeoff accidents described in Chapter 2. Overruns typically involve an unstable approach; the plane is generally too high or too fast over the runway threshold, and the plane lands too far and too fast down the runway. (By contrast, Asiana Flight 214 is an example of a "low and slow" approach, resulting in an underrun.) A frequent common denominator is that the flight crew fails to apply thrust reversers in a timely manner. Overruns often involve bad weather. The bad situation can still be salvaged with a go-around, that is, if an unstable approach presents potential for the plane landing too far down the runway and/or traveling too fast, the flight crew must quickly recognize the problem, apply maximum thrust, abort the landing, and consider other options (another approach or diversion). During the go-around, the aircraft may momentarily touch down, but at some point pilots are committed to the landing, as there is insufficient runway remaining to take off if the decision is delayed during the landing roll. Most airlines state that once reverse thrust has been actuated, the pilots are committed to the landing, though this varies somewhat by carrier and aircraft. All these factors were present for Air France Flight 358 on August 2, 2005.

A four-engine Airbus 340 with 309 passenger and crew attempted to land after an eight-hour flight from Paris. Airbus designates their flap extension settings as Flaps 1, 2, 3, or full. During the initial descent, Flap 2 was selected, the landing gear was extended, and the autopilot disengaged as the plane descended through 4,000 feet. Flap 3, then Flap full were selected; the plane converged on the glideslope as it descended through 3,000 feet and the autopilot was reengaged. The plane intercepted the glideslope about 8.7 miles from the runway threshold. Nothing about the approach to that point was remarkable.

The attempted landing occurred in a heavy downpour with lightning flashing all around the plane, providing crew distractions and making visibility (and stopping) more difficult. (Rainfall of nearly 4 inches per hour was recorded shortly before the accident.) The accident plane was flying at the target approach speed of 140 knots = V_{ref} + 5 knots. Just 35 seconds before touchdown, and passing through 323 feet, the autopilot and autothrust (called "autothrottle" by other manufacturers including Boeing; manufacturers frequently have different nomenclature for similar things) was disconnected. At about the same time, the wind shifted from a 90° crosswind at about 20 knots to a 10-knot tailwind. The pilot sensed excessive sink rate and advanced the throttles from 42% to 81% N1.

The target approach has the plane passing 50 feet above the runway threshold at V_{ref} = 1.23 × stall speed. The stall speed varies with flap setting and weight. V_{ref} for Flight 358 was 135 knots. Just 14 seconds before touchdown, Flight 358 crossed the threshold at a height of 70–80 feet and an indicated airspeed of 147 knots; because of increased thrust, it accelerated to 150 knots just 2 seconds later. The plane should have been decelerating as it crossed the threshold. The A340 flight manual states passing the threshold at the correct speed, but 50 feet too high adds 950 feet to the landing rollout.

While flying over the runway, there were repeated pitch and roll pilot inputs indicating a high pilot workload: the plane was moving too fast and the pilots were having trouble seeing the runway in the heavy downpour. The minor control and visibility problems distracted the pilots, and a go-around would have been their best course of action. Flight 358 touchdown occurred 3,800-3,900 feet down the 9,000-foot runway at a ground speed of 150 knots. Later, in a simulator, a pilot approximately repeated a similar scenario with similar results: excess thrust was added when the wind rapidly shifted to a tailwind. If the autothrust instead remained engaged, there was a more modest thrust increase and the simulated touchdown was within the first 2,000 feet of the runway.

The A340 certified landing distance on a dry, level runway is about 5,700 feet. The certified landing distance is based on actual test flights in which the plane crosses the threshold at V_{ref} and a height of 50 feet. The certified distance, allowing for some error from perfection, takes the actual landing distance and multiplies by 1.666. The certification test does not use reverse thrust; Captain Hedges explains on page 276 why planes might land without thrust reversers. For a wet runway, the certified distance is increased by 15%. Theoretically, with optimal conditions, the plane can land in a much shorter distance by not using up the safety factor and by using reverse thrust.

Estimated stopping distances for various scenarios for the accident flight are given in Table 9.2 for a dry and wet runway (less than 3 mm of water) with medium autobrake setting and Table 9.3 for a contaminated runway (approximately 6–7 mm of water) with manual braking. The distances given in Tables 9.2 and 9.3 must be added to the touchdown point (about 3,800 feet) and compared to the runway length of 9,000 feet. All values exceed the length of the runway. Hydroplaning, a condition when the tires float on a thin film of water (there is essentially more water than the tread can dissipate), was not a factor in the accident or considered in Table 9.3.

Table 9.2. Landing Distance for A340 Landing with Medium Autobrake Setting on a Wet Runway (Less Than 3 mm of Water)

Wind (knots)	Dry, No Thrust Reversers (feet)	Wet, No Thrust Reversers (feet)
0	5,419	5,829
10 (tailwind)	6,557	7,053

Table 9.3. Landing Distance for A340 Landing with Manual Braking on a Contaminated Runway (Approximately 6 to 7 mm of Water)

Wind (knots)	No Thrust Reversers (feet)	Four Thrust Reversers (feet)
0	8,780	7,883
5 (tailwind)	10,075	9,068
10 (tailwind)	11,388	10,249

On Flight 358, the spoilers (six on each wing) deployed automatically as designed after main landing gear touchdown. Maximum manual brakes were applied about 2.5 seconds after touchdown. Full reverse thrust was not applied until 16.4 seconds after touchdown, with about 2,200 feet of runway remaining. This can be compared to the certification assumption of 5.1 seconds until thrust reverser deployment.

After initial touchdown, the plane bounced at least three times. On each impact, passengers were bounced upward with arms and legs flailing. Nine passengers received serious injuries during impact. The plane left the runway with a ground speed of 86 knots and came to rest with a final impact in a ravine about 600 feet past the runway. The Captain's seat was detached from its base—the only seat so damaged. It is estimated that an impact must be at least 16 G to fracture the seat. (Most likely the front of the plane hits first and absorbs much of the impact, thus lowering G loads elsewhere.) The Captain, with back injuries, walked through the cabin looking for stragglers before leaving the aircraft.

The control tower activated the crash alarm 26 seconds after the plane left the runway. Firefighters witnessed the accident and arrived within one minute of the alarm. All 309 people on board safely evacuated ahead of the fire. The 297 passengers included three infants, one blind person, and three people in wheelchairs. Safety rules require one Flight Attendant for every 50 passenger seats at a minimum; for better service, there were 10 on board the aircraft. The extra Flight Attendants helped the expeditious evacuation.

The fire occurred in four areas: both wing main landing gear areas, fuselage door near the pressure bulkhead, and at the auxiliary power unit (tail section). Exits L3 and L4 were not opened because of fire. L2 was opened, but the slide failed to deploy. L2 was still used by 16 passengers who jumped 10-12 feet, with two people seriously injured. Only a few passengers exited from L1; the partially inflated slide was punctured. Attendants initially decided not to use R1 and R2 because of a nearby creek, but changed their minds as the smoke thickened, though only a few passengers exited R1 and R2. The R3 slide properly inflated but was punctured by debris. Exits L1, L2, L4, R1, R2, and R4 had double lane slides, but passengers tended to slide one at a time. The exits are shown in Figure 9.3.

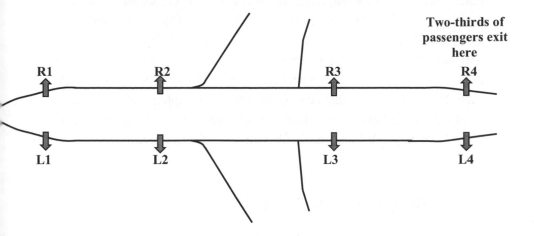

Figure 9.3. Layout of emergency exits on Flight 214.

Demonstrating the need for a hasty evacuation, the entire top half of the fuselage was quickly consumed in fire. Even with half of the exits unusable, it is estimated the evacuation was completed in a little over two minutes—a stellar achievement for the Flight Attendants.

Much of the interior planes are plastics made from petroleum. (Fire safety has improved tremendously in recent decades. Additives delay ignition and spread.) Though extensive fire damage prevented definitive fire forensics, fuel obviously spilt from the wing tanks and ignited. It's unclear whether ignition resulted from hot brakes or hot engine parts; certainly, an abundance of ignition sources were present. It's even possible brake hydraulic fluid ignited first.

Certification Evacuation Tests

Each new plane design must be evacuated with half the exits randomly blocked in less than 90 seconds—approximately the conditions of Flight 358. Testing is done with passengers carrying simulated babies and 35% must be over age 50. Attempting surprise, the passengers are seated for hours. Testing can be dangerous. In 1991, a 60-year-old woman was paralyzed after breaking her neck during certification testing for the McDonnell Douglas MD-11, which is about the same size as the Airbus A340 Toronto accident plane. In 2005, the Airbus A380 certification test occurred with 873 evacuating in 78 seconds with half the exits blocked. One person broke a leg. The A380 slides (which double as rafts and completed sea trials in the Pacific Ocean) inflate in 6 seconds. For additional safety, the fuselage must resist burn-through from a pool of burning fuel for five minutes.

Captain Hedges Explains: Energy Management and Overruns

Air France 358 is illustrative of the challenges pilots find in real-world bad-weather flying. After a long ocean crossing, the A340 descended into Toronto, where there were thunderstorms and abundant lightning. The crew wisely briefed procedures to follow in case they encountered wind shear and carefully deliberated whether to divert to an alternate airport (in this case Ottawa) because, owing to weather delays, they were close to reaching the minimum fuel needed to divert. While this was going on, the crew was also deviating around weather on the approach, configuring the aircraft for landing and going through briefings and checklists. The flight was cleared for an instrument landing system approach to runway 24L with the First Officer flying the aircraft. Weather for the approach varied dramatically from visual conditions to flight through dark clouds, heavy rain, and substantial turbulence.

During most of the approach, the autopilot and autothrust systems were engaged; at 323 feet above the ground, the First Officer disconnected the autopilot and shortly thereafter the autothrust. He then manually added thrust as airspeed decreased, and he perceived the aircraft sink rate was increasing (this could have signaled that dangerous wind shear was ahead). The wind conditions were rapidly changing from a headwind and crosswind to a 10-knot tailwind and crosswind, which increased the plane's ground speed. Because of all these rapidly changing factors, the A340 crossed the runway threshold about 40 feet higher than normal. This normally results in a touchdown farther down the runway than normal, as it takes more distance to lose more altitude at the same rate of descent. As the plane crossed the threshold, it entered an intense downpour that reduced the crew's visibility of the runway. Because the First Officer had added power to arrest the speed decay and sink rate, combined with the plane being higher than desired, Flight 358 touched down outside the touchdown zone about 3,800 feet down the runway, leaving only 5,100 feet to stop on wet pavement.

Although reverse thrust is helpful in slowing the aircraft, it has limitations and is not normally used in computing landing distances. An analysis of the flight control inputs made by the First Officer showed that he had his hands full and was working hard to control the aircraft with frequent, sometimes large, control inputs to the side-stick in both pitch and roll. After touchdown, he quickly applied maximum manual braking (more than the autobrakes can provide in this circumstance) and in the gusty wind conditions attempted to get the A340 to the runway centerline and keep it there. When the aircraft touched down, the aircraft heading was about 6° to the right of the runway heading owing to the crosswind component; this was in accordance with Air France and Airbus recommendations (this is called "crabbing" or "landing in a crab"). His judicious and delayed use of reverse thrust because of this mis-alignment between the aircraft and runway headings complied with Air France procedures, but further increased the landing distance. (The Captain failed to make standard landing callouts during the landing, which also likely delayed his use of reverse thrust.) Although this might seem counterintuitive, using reverse thrust with a slippery runway and a crosswind can lead to losing di-rectional control of the aircraft, as both the crosswind and reverse thrust vec-tors are attempting to make the aircraft depart from the side of the runway. Because the engines on an A340 are far from the aircraft centerline, there are large moment arms that can work to destabilize the plane; further, if a reverser does not deploy correctly, the resulting asymmetry may make the aircraft rap-idly depart the side of the runway if the pilot does not quickly respond to the situation. There are a lot of dynamic things happening in quick succession in a landing of this nature, and a great deal of training is devoted to dealing with convective weather, high winds, slippery runways, and stabilized approaches.

After the aircraft left the runway at about 80 knots, it crossed a ravine and came to rest with the cabin largely structurally intact. The pilots surveyed the cabin to ensure all occupants were safely evacuated before leaving the aircraft. Thanks to the excellent work of the Air France Flight Attendants, everyone on the aircraft evacuated safely before the plane was consumed in the fire after the crash. This shows the value of a well-trained cabin crew.

Of all the lessons learned from Air France 358, the most important for pilots has been emphasizing a stable approach all the way to the landing, and if a stable landing can't be ensured, to perform a go-around and either attempt another approach or divert to an airport with better weather. Airlines now have specific stabilized approach criteria that pilots must adhere to or go-around; if Air France 358 and Asiana 214 had elected to go around in a timely manner, this chapter would be completely different.

Epilogue

Although this is a book about plane crashes, it's important to view aviation safety in context. Many of the accidents in this book were spectacular and generated tremendous media coverage, but they are exceptionally rare. The Boeing 707 entered service in late 1958 and is now retired from the world's fleets. In its lifetime, it amassed a hull loss rate of 8.84 per million flights, a number that would be wholly unacceptable today but represented a huge improvement over the propeller-driven airliners that preceded it. Starting with the B-707, four generations of airliners have entered service, and with each generation, safety has improved. As an example, there have been three (soon to be four) generations of Boeing's popular 737 aircraft. The first generation (B-737-100/200) had a hull loss rate of 1.75 per million departures; for the next generation (B-737-300/400/500), the rate had improved to 0.57, while the newest iteration (B-737-600/700/800/900) is down to 0.23. Other aircraft families have generally improved similarly, although there have been outliers in both directions. The Lockheed L-1011 and McDonnell Douglas DC-10 were fierce competitors in the early wide-body market in the second generation of jet airliners, a group that also included the original B-747. Despite fulfilling the same mission, the L-1011 had an accident rate of 0.74 per million flights, while the DC-10 was four times worse, with a rate of 2.92 (for comparison, the early B-747's rate was 2.84). Airbus saw marked improvements in safety between their last conventional aircraft, the A310, and the first fly-by-wire airliner in service, the A320 family, with its enhanced automations and protections: the A310 had an accident rate of 2.31 per million flights, while the A320 family has seen just 0.26, a ninefold improvement. Subsequent designs have been even safer: there have been two A330 losses (for a rate of 0.27 per million flights) and two A340 losses, including the Air France overrun discussed in this book, yielding an accident rate of 0.66 per million flights. Despite the two A340 losses, the aircraft has never experienced a fatal accident. This trend has continued with all manufacturers: as of this writing, there have been no accidents involving the B-717, B-787, A380, or A350. The point of all this is to say that engineers have improved every generation of aircraft, and pilots have gotten better at flying these increasingly sophisticated machines.

From 1959 (the start of the jet age) until 2015, there were 1,905 airline accidents,[1] two-thirds of which (1,289) involved no fatalities. This gets to the point that most airline accidents are survivable, and with increasingly well-engineered cabins and structures, aircraft are more crashworthy than ever (consider both the Asiana and Air France landing accidents). In that same time frame, the world airliner fleet has flown 1,321,000,000 hours and had 713,000,000 departures. These are all reassuring numbers, but statistics can be used to imply significance when none is there, too.

Depending on one's viewpoint, 2014 was either the safest or the worst year for aviation safety in the past five years. How could that be? In 2014, there were seven fatal airliner accidents worldwide, which is the best rate in the past five years, but unfortunately two of those accidents were Malaysia Airlines Flight 370, which disappeared under suspicious circumstances, and Malaysia Airlines Flight 17, which was shot down with a missile, accounting for a disproportionately high number of casualties, bringing the fatalities for the industry for 2014 to 904, by far the highest in the last five years.

Aviation is a demanding profession, regardless of the actual role of any participant. Engineers and scientists are among the highest paid in industry and constantly work on design improvements for safety and efficiency, and have over time markedly improved aircraft in every way. New aircraft are carefully engineered with safety and ergonomics in mind: the aircraft are easier to fly in increasingly complex airspace and are easier to maintain, with fewer opportunities for errors engineered in at every step. Mechanics have become computer literate and have adapted to the increased integration between aircraft systems and diagnostic skills necessary to do their jobs in a prompt and safe manner to ensure aircraft are dispatched safely and on time. Airlines have become more safety focused and have been widespread adopters of safety management systems. Flight Attendants are regularly trained and tested on their reactions to various scenarios from smoke in the cabin to disruptive passengers, and they excel in evacuating passengers from an aircraft during emergencies with stunning speed and efficiency.

Regulatory agencies like the Federal Aviation Administration have changed the way they do business and now collaborate more with airlines and pilots in particular, with programs like the Aviation Safety Action Program (ASAP) that are not punitive, openly encourage pilots (and others) to admit errors they have seen or made, and offer suggestions to prevent a recurrence of that problem. That's not to say the FAA isn't still a tough regulator—it is—but the regulatory agencies have taken a more modern approach to safety systems, acknowledging that people make errors. In previous times, pilots would never admit to mistakes because the only possible outcome would be disciplinary action (up to and including permanent revocation of a pilot's license), but under ASAP, pilots who make a mistake while acting in good faith are encouraged to admit to what they have done so that the error can be shared and others can learn from it before the same error becomes a link in an accident chain. ASAP and other programs like it have been an unqualified success, and regulators, pilots, airlines, and unions all have universal praise for the system.

Pilots are still among the most highly trained and thoroughly tested employee groups anywhere. Pilot training involves intense dedication and study from the very first lesson, and it never stops. Pilots take various paths to the cockpit, but they all pay their dues and learn their lessons well in what is essentially a prolonged apprenticeship. Pilots may start in the military or go the civilian route, but the testing and training are rigorous at every step of the way; today, advanced simulation is ensuring that more and more scenarios are practiced before a pilot ever gets to fly the real aircraft. When a pilot is finally hired at a major airline (itself an arduous process), they will train to become First Officers. They are fully qualified to fly the aircraft and (in most countries) have an identical license to the Captain, but the Captain is by regulation the one person in command of the aircraft and is responsible for it. First Officers get the opportunity to fly every other flight in most operations alternating with the Captain, and it is a chance to learn decision-making skills, crew management, nuances about flying, and myriad other things as the hours accumulate in their logbooks. Ultimately, when their seniority allows it (at many airlines, they may be First Officers for 10–20 years before upgrade), they can bid to attend school to become a Captain, at which point they face an even more challenging curriculum and set of checkrides to make sure they are prepared for command. There is a high attrition rate early in a pilot's training (it is not uncommon for over half of any given class of students to wash out of military flight school), but by the time someone is ready to upgrade to Captain at a major airline, there is almost none: these are well-trained, experienced, highly motivated pilots, and failure is rare.

Are aircraft accidents still a threat? Absolutely. Fortunately, they are also uncommon, and everyone in the field works hard to keep it that way. The next time you step on an airliner, rest assured: you have a well-trained, experienced, and motivated crew that genuinely likes to fly. They want to get you your destination on time, but their first interest will always to be to get you there safely.

Notes

Preface

1. A check airman is an experienced pilot who can give other pilots recurrent examinations (check rides), normally in the simulator. A line check airman is a pilot with a great deal of experience in a given aircraft type who conducts training on the line during routine airline flights. Simulators are now so faithful to the real aircraft that for most airline pilots, the first time they ever fly the real aircraft is on a regularly scheduled flight. Line check airmen supervise these initial flights to make sure the pilots transitioning to a new plane don't have any questions and are comfortable flying the plane in real-world operations. Line check airmen are qualified to fly in either seat (in most normal operations, Captains fly from the left seat and First Officers fly from the right seat), as they conduct training for both Captains and First Officers and are subject to additional recurrent training and FAA observation. An FAA-designated pilot examiner is a pilot who is appointed by the FAA to administer both recurrent and initial check rides in a simulator (or aircraft) and oral knowledge examinations. A designated pilot examiner can issue a new certificate or type rating for a pilot completing a transition to a new aircraft.

Chapter 1 · Takeoff!

1. "Cargo Jet Explodes," *Newsday*, October 15, 2004.
2. "Jet Crash in Halifax Is Fourth for Airline," *Globe and Mail*, October 15, 2004.
3. "Cargo Jet Explodes."
4. "Jet Crashes on Takeoff in Halifax, Killing Seven," *Toronto Star*, October 15, 2004.
5. Comparable numbers for a heavy freight train are 0.2% fuel and 28% structure. A typical freight train replaces the cargo capacity of about 180 Boeing 747s. A realistic cruising thrust for the Boeing 747 is 45,000 lbs. For example, 45,000 lbs will pull about 30 million lbs of train on level ground. A train does not have to continuously consume power to lift itself.
6. Planes use ship navigational terminology. The nautical mile is one minute of arc (one-sixtieth of a degree), a distance historically subject to interpretation. By

international agreement, the nautical mile has been set to exactly 1,852 meters, or 1.150779 statute miles. A knot is one nautical mile per hour.

7. Pratt & Whitney and Rolls-Royce use EPR to set engine thrust. General Electric, the third major jet manufacturer, uses N1, which is the turbo-fan rotational speed.

8. A symmetric wing (i.e., a wing whose top surface is a mirror image of the bottom surface) has zero lift when the angle of attack (AOA) is zero degrees. A commercial airplane wing (not designed for flying upside down) has more curvature on the top surface and therefore a small lift force, even for zero AOA. The orientation that gives zero lift is actually a small negative AOA. Strictly speaking, the doubling or scaling of lift only works relative to the AOA with zero lift.

9. Northwest Airlines Flight 255, a McDonnell Douglas MD-82, crashed shortly after takeoff on August 16, 1987. The pilots failed to deploy the flaps and slats, and the alarms failed to alert the flight crew. A similar accident happened on August 31, 1988, to Delta Flight 1141, a Boeing 727.

10. The higher pressure below the wing curls around the wing tips into the lower pressure above the wing, creating a wake vortex, as shown in Figure 6.2 in Chapter 6. It takes energy to stir up all that air, effectively increasing the plane's drag. The proximity of the ground interrupts vortex formation. The net effect is higher lift and less drag.

11. The impact force can also be described with $f = ma$. Hit and stick is more abrupt with higher deceleration and corresponding higher-impact force. Hit and skip is less abrupt with lower deceleration and lower impact force.

12. Accident Investigation Board of Norway, *Report on the Serious Incident at Oslo Airport on 21 September 2004.*

Chapter 2 · Takeoff (Never Mind!)

1. In the United States, the NTSB has legal control of all transportation accident investigations, including those involving railroads, highways, pipelines, shipping, and aviation. The FAA creates and enforces all aviation safety rules. Accident investigations may result in the NTSB making safety recommendations to the FAA.

2. The kinetic theory of gases defines a gas to be a large number of small particles in constant random motion. The speed of sound of the gas also describes how fast a pressure pulse propagates through the gas. As the gas becomes hotter, the particles travel faster and correspondingly transmit a pressure pulse faster; hence hotter gases have a higher speed of sound.

3. Overspeed and burst can occur if the shaft breaks and decouples the input and output loads, or if control systems fail. More common is a manufacturing defect in one of the rotating disks. The defect initiates a fatigue crack that grows until the disk bursts, a topic covered at length in *Beyond the Black Box.*

4. The world's largest turbofan is the General Electric GE90-115B used on the

two-engine Boeing 777-300ER (ER stands for extended range; its first commercial flight was in 2004). The significantly larger Boeing 747 and Airbus A380 both use four smaller engines. The diameter of the GE90-115B, at 135 inches, is just 13 inches smaller than the diameter of the Boeing 737 fuselage. The GE90-115B has a pressure ratio of 42, a bypass ratio of 9, and a thrust rating of 115,300 lbs.

5. Just two minutes after takeoff at an altitude of 2,818 feet and airspeed of 185 knots, the Airbus A320 encountered a flock of Canada geese.

6. Dutch Safety Board, n.d.

7. The reverse thrust shown in Figure 2.10 is for the General Electric CF6-80C2 engines. These newer and more powerful engines have maximum forward static thrust of 59,000 lbs, slightly more than 53,000 lbs of rated thrust for the Pratt & Whitney engines on the Kalitta accident plane.

8. Air Accident Investigation Unit (2009).

9. The L-1011, with only 250 planes manufactured, competed with the McDonnell Douglas DC-10, Boeing 747, and Airbus A300. The L-1011 was the last commercial plane made by Lockheed. Today, Lockheed Martin only manufactures military planes.

10. The L-1011 departed Saudi Arabia on December 22, 1980. While climbing through 29,000 feet, the flight crew heard a loud explosion shortly before the cockpit door flew open. The Flight Attendant notified the crew there was a large hole in the cabin floor directly over the left main landing gear. A tire and wheel were visible through the hole. Metal fatigue caused failure of the wheel flange. The sudden and explosive release of the tire pressure projected part of the wheel rim through the bulkhead and another part into the cabin floor. Explosive release of the pressurized air in the fuselage ejected two children through the hole. Historically, the L-1011 had numerous wheel failures of little consequence. The problem was eventually solved by reinforcing the wheels.

11. It takes 100 foot-lbs of energy to lift 50 lbs 2 feet up.

12. The Concorde cut its tire on a metal strip during takeoff. The metal had fallen off a departing plane just five minutes before. A nearly 10 lb hunk of tire struck the Concorde's fuel tank at an estimated 310 mph. The pressure pulse ruptured the fuel tank. A massive fire structurally damaged the plane shortly after takeoff, killing all 113 on board. The economically challenged Concorde fleet was grounded for good in 2003. Because of the relative layout of tires and fuel tanks, this type of accident cannot happen on other modern jets.

13. "A Grisly Triptych of Disaster," *Time Magazine*, September 27, 1982.

14. In the 1960s, the rate of rejected takeoff incidents and accidents was 6.3 per 10,000,000 takeoffs. In the 1990s, the rate was down to 1.4 per 10,000,000 takeoffs.

15. All commercial aircraft have a minimum equipment list (MEL), which spells out what equipment can safely be inoperative on an airplane as well as the condi-

tions and restrictions associated with those deferrals. This is an important document, as otherwise an airplane would have to be perfect before every flight. Although perfection is always desired, it isn't always feasible. Imagine being a passenger on a flight and having it cancel because a window shade was broken or an overhead bin was inoperative. Though those are relatively trivial examples, the MEL is a huge book, and many of the items in it are more important. For example, an aircraft might be able to dispatch with its autoland system deferred by the MEL. That would likely restrict the lowest visibility the flight could land with, but if the weather at the destination was clear, autoland wouldn't be needed anyway.

Perhaps surprisingly, some aircraft can defer items that affect stopping distance using the MEL, and although this varies widely by specific aircraft type, examples can include the antiskid system of the brakes, automatic spoiler deployment, a thrust reverser, or even a brake or two. Why is this safe? Because the aircraft has been evaluated and certified for those situations, and each of them will come with restrictions (in this case weight penalties, which can be huge). Depending on the exact MEL and aircraft, the penalty might be thousands of pounds of reduction in maximum takeoff weight, and there might be other requirements as well (e.g., all thrust reversers must be operational if a brake is inoperative). Interestingly, one MEL that rarely causes performance penalties is for an inoperative thrust reverser, since on most planes the effects of reverse thrust are not accounted for in certification stopping distances (there are a few exceptions, however). The Captain has the right to refuse an aircraft for an MEL if he or she thinks it compromises safety.

Chapter 3 · Controlling the Plane

1. Republic of Cameroon, n.d.
2. Ibid.
3. Some autopilots, particularly in older aircraft, have two primary modes of operating with the autopilot engaged: "Command" and "Control Wheel Steering." Command is by far the more commonly used mode, as it will fly the aircraft in accordance with whatever settings the pilot has set on the autopilot controls. For example, if after takeoff a pilot has selected an altitude of 7,000 feet on the mode control panel, a mode of VNAV (vertical navigation), and a speed of 250 knots, the aircraft will accelerate to 250 knots, automatically trim the aircraft, direct the autothrottles to maintain climb power, and at 7,000 feet capture and hold the selected altitude and then retard the throttles to maintain 250 knots. During all this, the FMA will tell the pilot exactly what modes are operational for the autopilot, autothrottles, speed, and vertical and lateral navigational modes. It is critical that both pilots (particularly the Pilot Flying) are aware of these modes to ensure that the aircraft is doing what it is intended to do. The actual nomenclature displayed on the FMA differs by aircraft model but is functionally similar and always thoroughly covered in training.

Control Wheel Steering (CWS) is an uncommon mode, and many airlines prohibit its use even when installed. Most new aircraft no longer come with CWS, but because the original design is now 50 years old, the B-737 still does. CWS is essentially a stabilization mode. If an autopilot is engaged in CWS, the aircraft will trim for the commanded speed and will generally hold whatever pitch and roll angles the pilots direct within limits. The key difference here is that the pilots must still manually fly the aircraft with the control yoke: CWS trims and holds whatever attitude the pilots set with the yoke. This only marginally reduces workload and has few practical uses and several practical pitfalls, which is why most planes no longer have this mode, and when it is installed, even fewer pilots use it.

4. The famous barrel roll occurred in front of 200,000 spectators (and a reviewing stand packed with aviation executives) turned out for boat races on Lake Washington in Seattle. Upon seeing his company's biggest investment upside down, Bill Allen, president of Boeing, "looked like a clinical example of apoplexy."

Chapter 4 · Vanished!

1. In 2005, a hypoxic crew flew a Boeing 737 on autopilot for almost three hours before crashing in Greece. See Bibel, *Beyond the Black Box*.

2. New lithium batteries had to be designed and pass certification tests, perhaps requiring a few design cycles. Long-term contracts for existing batteries were broken with potential legal consequences. Existing supply chains were disrupted when the new and larger lithium batteries were judged a hazardous material. And, finally, planes had to be taken out of service to install the new batteries.

3. Centrifugal force is an inertia effect and not a force.

4. One such limit is flutter, a type of structural vibration with several mechanisms. The wing's motion interacts with the aerodynamic forces in such a way that magnification occurs, with potentially catastrophic results. The wings, stabilizers, or any of the plane's control surfaces can flutter. High-speed buffet can also cause flutter. Flutter is avoided by not exceeding critical speeds, and it is fundamental to safe flight. Flutter is a solved engineering problem with no serious occurrences in the modern jet era.

5. The pitot tube measures total pressure, which consists of static pressure plus dynamic pressure (a.k.a. ram pressure). Static pressure is the pressure of still air. Static pressure is also the weight of a column of air above a 1-inch square. Indicated airspeed depends on the total pressure minus the static pressure. The Airbus A330 has separate static and dynamic pressure ports.

6. The simulators are calibrated with test flights. Only limited stalling occurs during certification flights. Stalling variables are weight, altitude, location of center of gravity, and bank angle.

Chapter 5 · Practice Makes Perfect

1. The MD-80 series is an updated version of the McDonnell Douglas DC-9. Nearly 1,200 were produced between 1979 and 1999.

Chapter 6 · Turbulence

1. Fong and Mathieu, "15 Seconds of Terror."
2. Chaney et al., "15 Seconds of Terror."
3. Fong and Mathieu, "15 Seconds of Terror."
4. Ibid.
5. Ibid.
6. National Transportation Safety Board, *Aircraft Accident Report*.
7. Rosenkrans, "Outmaneuvered Airflow."
8. Ibid.
9. National Transportation Safety Board, *Aircraft Accident Report*, 19.
10. Ibid., 20.
11. Ibid.
12. Tex Johnston famously rolled the prototype for the Boeing 707 upside down in 1954.
13. Thunderstorms remain a danger to inexperienced small-plane pilots flying in the dark without weather radar.
14. The last major wind shear accident in the United States occurred on August 2, 1985, when a Lockheed L-1011 crashed during approach into Dallas, killing 134 of the 163 occupants. Elsewhere in the world, airplanes continue to crash in wind shear, often after incorrect decision making about diverting to another airport and failing to follow standard procedures to avoid active weather cells.
15. An accident is defined by substantial damage to the plane (rare for turbulence) or a serious injury.

Chapter 7 · The 168-Ton Glider

1. Fuel is often moved around for economic reasons; this is called tankering. Fuel is more expensive in Portugal than Canada, for instance.
2. The real meaning of this regulation, known as ETOPS (for extended operations, formerly known as extended-range twin-engine operational performance standards), is to limit the plane from flying no farther than 150 minutes of flight from a safe location to land per the airline's operation specifications. Different aircraft and airlines have varying ETOPS certifications for different ranges between suitable alternates. The current record holder is the Airbus A350, which has ETOPS 370, meaning it can be 370 minutes from a suitable alternate. Originally, ETOPS only covered two-engine aircraft, but in 2008 the Federal Aviation Administration

published requirements for aircraft with three or four engines to also comply with ETOPS requirements, although the applicability is slightly different in the details.

3. The airline's engine controller interfaces directly with Rolls-Royce and coordinates off-wing maintenance of engines.

4. Maximum landing weight for a standard A330-200 is 396,800 lbs, while the maximum takeoff weight (MTOW) in the Air Transat configuration is 473,000 lbs. The huge difference between the two weights is consumed fuel. In emergencies, some aircraft (including many A330s) have fuel jettisoning systems installed to help in rapidly decreasing aircraft weight. In cases of extreme emergency, all aircraft can be landed at up to their MTOW at a sink rate of 6 ft/sec without structural damage; and at up to 10 ft/sec without damage at the lower maximum landing weight.

Chapter 8 · Approach

1. "Passengers Tell of Their Brush with Death."

2. Historically, fuel tank vapors have exploded when ignited by electrical arcing (caused by aging or damaged wiring, not a crash event)—a problem now solved by fuel tank inerting. Fuel vapors can also self-ignite for temperatures above 400°F. Hot air rises and sucks in colder air at the bottom, which is why fires burn up and why only the top half of the fuselage burns away. A fire can burn up the entire fuselage from a pool of spilt fuel, a problem that apparently has not resulted in a fuel tank explosion since at least the 1980s. (The National Transportation Safety Board lists major foreign mechanical failures. There are no fuel tank explosions in the NTSB database except for electrical arcing.) Engines and landing gear that break away cleanly to minimize fuel tank damage apparently do work.

3. In 1998, a fire aboard Swissair Flight 111 disrupted power in the cockpit, leading to a crash. The flammability test for the Mylar moisture barrier was flawed. A strip placed above a Bunsen burner passed the test by shriveling up and away from the flame. Mylar was subsequently removed from all planes. The Concorde disaster of 2000 was briefly discussed in Chapter 2.

4. Both are explained with the kinetic theory of gas. The impact of individual air molecules creates pressure on the liquid surface. This pressure must be overcome by the liquid molecules trying to jump into the vapor space. Lower pressure makes boiling easier, and higher pressure the opposite.

Epilogue

1. This discussion considers only certified jet airliners with a gross weight of at least 60,000 lbs. A hull loss is an accident in which an aircraft is either destroyed, damaged beyond economical repair, or missing.

Chapter 1

Accident Investigation Board of Norway. *Report on the Serious Incident at Oslo Airport Gardenmoen on 21 September 2004 Involving Flight KAL520 Boeing 747-400F Registered HL7467 Operated by Korean Air.* Lillestrom: Accident Investigation Board of Norway, March 2007.

Air Accidents Investigation Branch. *Serious Incident, Airbus A330-243, G-OJMC, 28 October 2008.* London: Air Accidents Investigation Branch, November 2009.

Airbus. *A380 Technical Training Manual Maintenance Course: T1 & T2 (RR/Metric) Level I—ATA 31 Indicating/Recording Systems.* Toulouse, France: Airbus, March 21, 2006.

Airbus Customer Services. *Flight Operations Briefing Notes: Preventing Tailstrike at Takeoff.* Toulouse, France: Airbus, November 2005.

Aircraft and Railway Accidents Investigation Commission. *Nippon Cargo Airlines Flight 62: Boeing 747-200F. New Tokyo International Airport, Narita, Japan, October 22, 2003.* Tokyo: Aircraft and Railway Accidents Investigation Commission, July 30, 2004.

Anderson, David F., and Scott Eberhardt. *Understanding Flight,* 2nd ed. New York: McGraw Hill, 2010.

Associated Press. "Cargo Jet Explodes." *Newsday,* October 15, 2004.

Australian Transport Safety Bureau. *Tailstrike and Runway Overrun Melbourne Airport, Victoria: 20 March 2009.* AO-2009-12. Canberra, ACT: Australian Transport Safety Bureau, 2011.

———. *Take-Off Performance Calculation and Entry Errors: A Global Perspective.* AR-20090052. Canberra, ACT: Australian Transport Safety Bureau, 2011.

Bibel, George. *Beyond the Black Box: The Forensics of Airplane Crashes.* Baltimore: Johns Hopkins University Press, 2008.

Boeing Company. *747 Airplane Characteristics.* D6-58326. Chicago: Boeing Company, May 1984.

———. *Boeing 747 Aircraft Maintenance Manual.* Chicago: Boeing Company, April 25, 2007.

――――. *Submission for National Airlines (NAL) 747-400BCF N949CA Takeoff Accident Bagram AFB, Afghanistan: 29 April 2013.* Chicago: Boeing Company, March 19, 2015.

Brotak, Ed. "Into Thin Air." *AeroSafety World*, November 2013.

Carbaugh, Dave. "Tail Strikes: Prevention." *AERO Magazine*, Quarter 1.07, 2007.

Carbaugh, Dave, and Linda Orlandy. "Avoiding Tail Strikes." *AERO Magazine*, Quarter 4.13, 2013.

Cathay Pacific Airways. *Boeing 747 Aircraft Maintenance Manual.* Hong Kong: Cathay Pacific Airways, 1989.

Chiles, Patrick. "Filling the Envelope." *Aviation Safety World*, December 2006.

Chung, Andrew, and Dale Brazao. "Jet Crashes on Takeoff in Halifax, Killing Seven; 'Just Wasn't Enough Runway.'" *Toronto Star*, October 15, 2004.

Dooley, Richard. "Jet's Engines Replaced Just before Crash in Halifax." *Ottawa Citizen*, October 17, 2004.

Dutch Safety Board. *Tail Strike during Take-Off: Boeing 737-800, Registration PH-HZB, Rotterdam Airport, on January 12, 2003.* The Hague: Dutch Safety Board, November 2006.

Eisler, Peter. "The Truck the Pentagon Wants and the Firm That Makes It." *USA Today*, August 1, 2007.

Federal Aviation Administration. *Aviation Maintenance Technician Handbook General: FAA-H-8083-30.* Washington, DC: FAA, 2008.

――――. *Type Certificate Data Sheet A16WE: Boeing 737 Series.* Washington, DC: FAA, August 24, 2009.

――――. *Type Certificate Data Sheet No. A20WE: Boeing 747 Series.* Washington, DC: FAA, December 10, 2013.

Foot, Richard. "747 Crashes, Killing All 7 Crew." *National Post*, October 15, 2004.

Gadher, Dipesh. "Overloading Blamed for Plane Crash That Killed Four Britons." *Sunday Times*, October 24, 2004.

Hanke, C. Rodney, and Donald R. Nordwall. *The Simulation of a Jumbo Jet Transport Aircraft.* Vol. 2, *Modeling Data.* Chicago: Boeing Company, September 1970.

Hradecky, Simon. "Crash: National Air Cargo B744 at Bagram on Apr 29th 2013, Lost Height Shortly after Takeoff Following Load Shift and Stall." Aviation Herald, June 4, 2013.

Jane's Information Group. *Jane's All the World's Aircraft 1999–2000.* London: Jane's Information Group, 1999.

"Jet Data Recorder Found at Site; Heat Damage to Box 'Substantial': 7 Killed in Cargo Plane Takeoff." *Toronto Star*, October 18, 2004.

Jones, Todd, et al. *Statistical Data for the Boeing-747-400 Aircraft in Commercial Operations.* DOT/FAA/AR-04/44. Washington, DC: US Department of Transportation/FAA, Office of Aviation Research, January 2005.

Macleod, Steve. "Cockpit Voice Recorder Destroyed." *Globe and Mail*, October 20, 2004.

——. "Takeoff Speed Too Low for Doomed Jetliner." *Globe and Mail*, October 23, 2004.

National Transportation Safety Board. *Allegheny Airlines, Douglas DC-9, Philadelphia, Pennsylvania, June 23, 1976*. NTSB-AAR-78-2. Washington, DC: NTSB, January 19, 1978.

——. *Northwest Airlines, McDonnell Douglas DC-9-82, N312RC Detroit Metropolitan Wayne County Airport, Romulus, Michigan, August 16, 1987*. NTSB/AAR-88/05. Washington, DC: NTSB, May 10, 1988.

——. *Delta Air Lines, Boeing 727-232, N473DA Dallas-Fort Worth International Airport, Texas, August 31, 1988*. NTSB/AAR-89/04. Washington, DC: NTSB, September 26, 1989.

——. *Aircraft Accident Report: Loss of Pitch Control During Takeoff—Air Midwest Flight 5481, Raytheon (Beechcraft) 1900D, N233YV, Charlotte, North Carolina, January 8, 2003*. NTSB/AAR 04/01. Washington, DC: NTSB, February 26, 2004.

——. *Descent below Visual Glidepath and Impact with Seawall Asiana Airlines Flight 214 Boeing 777 San Francisco, California, July 6, 2013*. NTSB/AAR-14/01. Washington, DC: NTSB, June 24, 2014.

——. *Operational Factors: Attachment 22—Witness Statements*. Washington, DC: NTSB, October 2014.

——. *Operational Factors: Attachment 23—Ypsilanti Simulator Work*. DCA13M1081. Washington, DC: NTSB, February 2, 2015.

——. *Steep Climb and Uncontrolled Descent during Takeoff National Air Cargo, Inc., dba National Airlines Boeing 747 400 BCF, N949CA: Bagram, Afghanistan, April 29, 2013*. NTSB/AAR-15/01. Washington, DC: NTSB, July 14, 2015.

"Overloading Ruled Out in Halifax Cargo Jet Crash." *Globe and Mail*, November 15, 2004.

Richer, Shawna. "Jet Crash in Halifax is Fourth for Airline." *Globe and Mail*, October 15, 2004.

——. "Why Jet's Tail Hit Runway Key to Crash Investigation." *Globe and Mail*, October 16, 2004.

Saarlas, Maido. *Aircraft Performance*. Hoboken, NJ: John Wiley. 2007.

South African Civil Aviation Authority. *Incident Report on 11 March 2003*. Ref. 0263. Johannesburg: South African Civil Aviation Authority, 2003.

Swedish Accident Investigation Board. *Uncommanded Rotation, Incident Involving Aircraft LN-RPL at Gothenburg/Landvetter Airport, O County, Sweden, on 7 December 2003*. Stockholm: Swedish Accident Investigation Board, 2006.

Ting, Dennis. *Reduced and Derated Thrust*. Chicago: Boeing Company, September 2009.

Transport Accident Investigation Commission. *Report 03-003 Tail Strike During Takeoff 12 March 2003*. Wellington, New Zealand: Transport Accident Investigation Commission, November 19, 2003.

Transportation Safety Board of Canada. *Reduced Power at Take-Off and Collision with Terrain: MK Airlines, 14 October 2004*. Gatineau: Transportation Safety Board of Canada, April 6, 2006.

Trofimov, Yaroslav, and Dion Nissenbaum. "U.S. Cargo Plane Crashes in Afghanistan, Killing 7." *Wall Street Journal*, April 30, 2013.

Webster, Ben, and Jan Raath. "One Airline, 4 Crashes, 8 Dead." *The Times*, November 20, 2004.

Williamson, A. M., and A. Feyer. "Moderated Sleep Deprivation Produces Impairments in Cognitive and Motor Performance Equivalent to Legally Prescribed Levels of Alcohol Intoxication. *Occupational & Environmental Medicine* 57, no. 10 (2000): http://dx.doi.org/10.1136/oem.57.10.649.

"Wrong Throttle Setting Cited in Probe of Cargo Jet Crash." *Globe and Mail*. December 17, 2004.

Chapter 2

Air Accident Investigation Unit. *Final Report on the Accident Occurred on 25 May 2008 at Brussels Airport on a Boeing B747-209F Registered N704CK*. Brussels: Air Accident Investigation Unit, July 10, 2009.

Airbus. *Takeoff and Departure Operations: Revisiting the "Stop or Go" Decision*. Toulouse: Airbus, December 2005.

Allen, Tim. "Operational Advantages of Carbon Brakes." *AERO Magazine*. Quarter 3.09, 2009.

American Association of State Highway and Transportation Officials. *Guide for Pavement Friction*. Washington, DC: American Association of State Highway and Transportation Officials, 2008

Australian Transport Safety Bureau. *Rejected Takeoff, Brisbane Airport, Qld—19 March 2006, VH-QPB Airbus A330-303*. Canberra, ACT: Australian Transport Safety Bureau, February 5, 2008.

Bardou, George. "Carbon Brakes." *FAST Airbus Technical Digest*, January 1987.

Bibel, George. *Beyond the Black Box: The Forensics of Airplane Crashes*. Baltimore: Johns Hopkins University Press, 2008.

Blake, Walt. *Jet Transport Performance Methods*. Chicago: Boeing Company, March 2009.

Boeing Company. *747 Airplane Characteristics Airport Planning*. D6-58326. Chicago: Boeing Company, May 1984.

———. *Takeoff Performance*. Chicago: Boeing Company, 2009.

Bureau Enquetes Accidents. *Accident on 25 July 2000 at La Patte d'Oie in Gonesse (95) to* the Concorde registered F-BTSC Operated by Air France. Report Translation. Le Bourget: Bureau Enquetes Accidents, January 16, 2002.

Choi, Bernard, and Joe Parke. Boeing 747-8 Performs Ultimate Rejected Takeoff. Chicago: Boeing Company, January 2013. http://www.boeing.com/Features/2011/08/bca_747-8FT_08_03_11.html.

Civil Aviation Accident Investigation Commission of Spain. Accident Occurred on September 13th, 1982, to McDonnell Douglas DC-10-30-CF aircraft EC-DEG, at Malaga Airport. Madrid: Civil Aviation Accident Investigation Commission of Spain, 1983.

Clark, Stephen F. "787 Propulsion System." AERO Magazine, Quarter 3.12, 2012.

Currie, Andy I. "Thrust Reverser System." FAST Airbus Technical Digest, July 1997.

Di Santo, Guy. "Proper Operation of Carbon Brakes." Paper presented at the 11th Airbus Performance and Operations Conference, Jalisco, Mexico, March 2001.

Dornheim, Michael A. "Passing the Test—The Hard Way." Aviation Week & Space Technology, March 11, 2002.

Dutch Safety Board. Rejected Takeoff after the Takeoff Decision Speed 'V1,' Boeing B737-800, at Eindhoven Airport 4 June 2010. The Hague: Dutch Safety Board, n.d.

Fabre, Cyril. "The Airbus High Tyre Pressure Test." FAST Airbus Technical Digest, August 2011.

Federal Aviation Administration. High-Speed Tire Maintenance and Operational Practices. AC 20-91A. Washington, DC: FAA, May 13, 1987.

———. Takeoff Safety Training Aid. Washington, DC: FAA, September 12, 1994.

———. Aircraft Tires. TSO-C62e. Washington, DC: FAA, September 29, 2006.

———. Flight Test Guide for Certification of Transport Category Airplanes. AC 25-7B. Washington, DC: FAA, March 29, 2011.

———. Aviation Maintenance Technician Handbook. Vol. 2, Airframe. FAA-H-8083-31. Washington, DC: FAA, 2012.

———. Type Certificate Data Sheet A22WE. Washington, DC: FAA, April 11, 2012.

Flight Safety Foundation. Tire Failure on Takeoff Sets Stage for Fatal Inflight Fire and Crash. Alexandria, VA: Flight Safety Foundation, September 1993.

GE Aviation. "The CF6 Engine." Accessed November 21, 2014. https://www.geaviation.com/commercial/engines/cf6-engine.

Goodyear Tire & Rubber Company. Global Aviation Tires. Akron, OH: Goodyear Tire & Rubber Company, January 2015.

Huber, Mark. "The Detroit Airlift." Smithsonian Air & Space Magazine, July 2001.

International Civil Aviation Organisation. ICAO Aircraft Accident Digest 1980: Lockheed L-1011, Accident over International Waters Near the State of Qatar on 22 December 1980. Montreal: ICAO, 1981.

Kalitta Air. "Our History." Accessed May 26, 2017. http://www.kalittaair.com/our-history/.

Koff, Bernard L. "Gas Turbine Technology Evolution—A Designer's Perspective." Paper presented at AIAA/ICAS International Air and Space Symposium and Exposition, Dayton, OH, July 14–17, 2003.

Mackness, Bob. "Brake Temperature." *AERO Magazine*, Quarter 1.02, 2002.

Merrit, D. R., and F. Weinhaus. "The Pressure Curve for a Rubber Balloon." *American Journal of Physics* 46, no. 976 (October 1978): http://dx.doi.org/10.1119/1.11486.

National Transportation Safety Board. *Brake Performance of the McDonnell Douglas DC-10-30/40 During High Speed, High Energy Rejected Takeoff.* Washington, DC: NTSB, February 27, 1990.

———. *Runway Overruns Following High Speed Rejected Takeoffs.* NTSB/SIR-90/02. Washington, DC: NTSB, February 27, 1990.

———. *Accident Report LAX97FA276.* Washington, DC: NTSB, February 11, 2000.

———. *Accident Report MIA04IA056.* Washington, DC: NTSB, December 20, 2005.

———. *Loss of Thrust in Both Engines after Encountering a Flock of Birds and Subsequent Ditching on the Hudson River US Airways Flight 1549 Airbus A320-214, N106US Weehawken, New Jersey, January 15, 2009.* NTSB/AAR-10/03. Washington, DC: NTSB, May 4, 2010.

Pratt & Whitney. "JT9D Engine." Accessed November 14, 2014. http://www.pw.utc.com/JT9D_Engine.

———. "J57 (JT3) Engine." Accessed November 22, 2014. http://www.pw.utc.com/J57_JT3_Engine.

Root, Rob. *Brake Energy Considerations in Flight Operations.* Chicago: Boeing Company, September 2003.

Sheehan, Jerry R. "Commercialization and Transfer of Technology in the U.S. Jet Aircraft Engine Industry." M.S. thesis, Massachusetts Institute of Technology, June 1991.

Stockton, William. "Blast Blamed in Mexican Crash." *New York Times*, April 10, 1986.

"Tire Blamed in Mexico Crash." *New York Times.* May 23, 1986.

Turbofan Engine Malfunction Recognition and Response Working Group. "Understanding Airplane Turbofan Engine Operation Helps Flight Crews Respond to Malfunctions." *Flight Safety Digest* 20, no. 3, March 2001.

UTC Aerospace Systems. *Goodrich 787 Electro-Mechanical Brake.* Troy, OH: UTC Aerospace Systems, May 2013.

Van Cott, H., and R. Kinbade. *Human Engineering Guide to Equipment Design.* Washington, DC: American Institute of Research, 1972.

van Es, G. W. H. *Rejecting a Takeoff after V_1 . . . Why Does It (Still) Happen?* NLR-TP-2010-177. Amsterdam: NLR Air Transport Safety Institute. April 2010.

Walus, Konrad J., and Zbigniew Olszwski. "Analysis of Tire-Road Contact under Winter Conditions." Paper presented at the World Congress on Engineering, London, July 6–8, 2011.

Wilhelm, Steve. "Mindful of Rivals, Boeing Keeps Tinkering with Its 737." *Puget Sound Business Journal*, August 8, 2002.

Witkin, Richard. "Sound and Fury over Jets: Port Authority's Delay in Accepting Boeing 707's Noise Level Has Clouded Start of Trans-Atlantic Service." *New York Times*, September 21, 1958.

Chapter 3

Baron, Robert I. "The Toxic Captain." *AeroSafety World*, March 2012.

Bibel, George. *Beyond the Black Box: The Forensics of Airplane Crashes*. Baltimore: Johns Hopkins University Press, 2008.

"Britons Feared Dead in Cameroon Plane Crash." *The Telegraph*, May 5, 2007. http://www.telegraph.co.uk/news/worldnews/1550687/Britons-feared-dead-in-Cameroon-plane-crash.html.

Doyle, Andrew. "Airbus to Reinforce Part of A380 Wing." *Flight International*, May 23, 2006.

"Engine Trouble Focus of Jet Crash." *Los Angeles Times*, May 8, 2007.Jedick, Rocky. "Spatial Disorientation." Go Flight Med, April 1, 2013. http://goflightmedicine.com/spatial-disorientation/.

Jotischky, Tim. "The Tragedy of Flight KQ507." *The Telegraph*, March 18, 2012.

Kaminski-Morrow, David. "Investigators Trawl Swamps for Clues on Kenya 737 Crash." *Flight International*, May 15, 2007.

———. "Kenya 737 Crash: Poor Airmanship Led to Disorientation." FlightGlobal, April28,2010.https://www.flightglobal.com/news/articles/kenya-737-crash-poor-airmanship-led-to-disorientation-341192/.

"Kenya Air Jet Goes Down in Cameroon." *New York Times*, May 6, 2007.

Lacagnina, Mark. "Beyond Redemption." *AeroSafety World*, August 2010.

McCrummen, Stephanie. "Passenger Jet Disappears over Cameroon." *Washington Post*, May 6, 2007.

Republic of Cameroon. *Technical Investigation into the Accident of the B737-800 Registration 5Y-KYA Operated by Kenya Airways that Occurred on the 5th of May 2007 in Douala*. Yaoundé: Republic of Cameroon, n.d.

Wall, Robert. "Snap to Attention." *Aviation Week & Space Technology*, February 20, 2006.

Chapter 4

"An Exceptional Project." *Fugro Cross Section Magazine*, May 24, 2015.

Associated Press. "3 Pieces of Evidence Point to Jet's Takeover." *Seattle Times*, March 16, 2014.

Bibel, George. *Beyond the Black Box: The Forensics of Airplane Crashes*. Baltimore: Johns Hopkins University Press, 2008.

Blake, Walt. *Jet Transport Performance Methods*. Chicago: Boeing Company, March 2009.

Bremner, Charles. "Airliner Vanishes in Tropical Thunderstorms." *The Times*, June 2, 2009.

Brennan, Zoe. "Why Other Planes Could Fall Out of the Sky." *Daily Mail*, May 29, 2010.

Brothers, Caroline. "Search at Sea for Answers in Loss of Jet Signs of Electrical Fault on Airbus Carrying 228 from Brazil to France." *International Herald Tribune*, June 2, 2009.

Bureau d'Enquêtes et d'Analyses. *Air France 447, A-330-203, F-GZCP Final Report, Near the TASIL Point, in International Waters, Atlantic Ocean, 1st June 2009*. Paris: Bureau d'Enquêtes et d'Analyses, July 2012.

———. *Sea Search Operations*. Paris: Bureau d'Enque_tes et d'Analyses, October 2012.

Campbell, Matthew, and Richard Woods. "Crash Jet 'Split in Two at High Altitude.'" *Sunday Times*, June 14, 2009.

Clark, Nicola. "French Jet Hit Ocean Still Intact, Officials Say." *New York Times*, July 3, 2009.

Clinch, Philip. "Flight Tracking Equipage Gaps—ADS B or C?" Air Traffic Management. November 11, 2015. http://www.airtrafficmanagement.net/2015/11/31055/.

Croft, John. "Never Again." *Aviation Week & Space Technology*, August 4, 2014.

———. "Regaining Control." *Aviation Week & Space Technology*, December 7, 2015.

Diehl, Alan E. "Why It's Time to Re-Evaluate MH370 Search Efforts." *Aviation Week & Space* Technology, August 31, 2015.

Edwards, Cody. "Ocean Debris Is Not from Air France Flight 447." *Washington Post*, June 6, 2009.

Escher, Federico, and Bradley Brooks. "Air France Jet Likely Broke up in Mid-Air." *Globe and Mail*, June 4, 2009.

Esler, David. "Global Advance of ADS-B." *Aviation Week & Space Technology*, December 11, 2015.

"Flight Tracking Initiatives & Systems." *Aircraft Commerce*, June/July 2015.

Flynn, Ed. *Understanding ACARS*, 3rd ed. Reynoldsburg, OH: Universal Radio Research, 1995.

Fricker, Martin. "'It's Highly Likely a Bomb Went Off'" *Daily Mirror*, June 4, 2009.

Heico. "Dukane Seacom Creates First Certified 90 Day Underwater Locator Beacon." News release, June 23, 2015.

International Civil Aviation Organisation. *North Atlantic Operations and Airspace Manual*. Montreal: ICAO, 2014/2015.

Krahe, Chris. "Airbus Fly-By-Wire Aircraft at a Glance." *FAST Airbus Technical Digest*, December 1996.

Ministry of Transport Malaysia. *Factual Information Safety Investigation for MH 370 Malaysia Airlines MH370 Boeing B777-200ER 08 March 2014*. Putrajaya: Ministry of Transport Malaysia, 2015.

National Transportation Safety Board. "Minutes of the Emerging Flight Data and Locator Technology Forum." Washington, DC: NTSB, October 7, 2014.

"Operator Benefits of Future Air Navigation System (FANS)." *AERO Magazine*, Quarter 2.98, 1998.

Palopo, Kee, et al. "Wind-Optimal Routing in the National Airspace System." Paper presented at the 9th AIAA Aviation Technology, Integration and Operations Conference, Hilton Head, SC, September 21–23, 2009.

"Public Trust, Rapid Response." *ICAO Journal* 1, 2015.

Richards, William R., et al. "New Air Traffic Surveillance Technology." *AERO Magazine*, Quarter 2.10, 2010.

Rockwell Collins. "High Frequency Data Link (HFDL) Frequently Asked Questions." Accessed January 14, 2016, http://archive.is/wBKAZ.

Schroeder, Jeffery A., et al. "An Evaluation of Several Stall Models for Commercial Transport Training." Paper presented at the AIAA Modeling and Simulation Technologies Conference, Atlanta, GA, June 16–20, 2014.

Tomayko, James E. *Computers Take Flight: A History of NASA's Pioneering Digital Fly-By-Wire Project*. Washington, DC: National Aeronautics and Space Administration History Office, 2000.

Turner, Aimee. "Radio Spectrum Win for Global Flight Tracking." November 11, 2015. http://www.airtrafficmanagement.net/2015/11radio-spectrum-allocated-for-global-flight-tracking/.

Van Wagenen, Juliet, and Caleb Henry. "ITU Flight Tracking Spectrum Allocation." *Avionics Magazine*, November 13, 2015.

Wald, Matthew L. "New Signs That Plane Broke Up in Flight." *New York Times*, June 11, 2009.

Chapter 5

Allerton, David. *Principles of Flight Simulation*. Chichester, UK: John Wiley, 2009.

American Society of Mechanical Engineers. *The Link Flight Trainer*. New York: American Society of Mechanical Engineers, May 2000.

Bauman, Richard. "Link to the Future." *Aviation History*, May 1, 2014.

Bibel, George. *Beyond the Black Box: The Forensics of Airplane Crashes*. Baltimore: Johns Hopkins University Press, 2008.

——. *Train Wreck: The Forensics of Rail Disasters*. Baltimore: Johns Hopkins University Press, 2012.

Brooks, Don. "Engineering Simulators Enhance 777 Development." *Aerospace America*, August 1993.

Christensen, Brad. "Full Motion Flight Simulator in the Classroom." *TechDirections*, February 2005.

Elson, Benjamin M. "Boeing Gains Real-Time Flight Data." *Aviation Week & Space Technology*, January 17, 1983.

Federal Aviation Administration. *Aircraft Simulator and Visual System Evaluation and Approval*. AC 121-14C. Washington, DC: FAA, August 29, 1980.

Frisbee, John L. "Valor: AACMO—Fiasco or Victory?" *Air Force Magazine*, March 1995.

Hewett, Mike. "The 737 Virtual Flight Deck." *AERO Magazine*, Quarter 4.98, 1998.

"Keeping the Complex Simple." *Flight International*, February 13, 1988.

Lapiska, Carl, et al. "Flight Simulation: An Overview." *Aerospace America*, August 1993.

Lee, Alfred T. *Flight Simulation*. Farnham, UK: Ashgate, 2005.

NASA Space Science Data Coordinated Archive. "Mariner 4." Accessed July 15, 2014. http://nssdc.gsfc.nasa.gov/nmc/spacecraftDisplay.do?id=1964-077A.

"Pilot Transition Tested Using Simulators Alone." *Aviation Week & Space Technology*, March 13, 1978.

Riordan, Michael. "The Incredible Shrinking Transistor." *MIT Technology Review*, November 1, 1997.

Schilling, Lawrence J., and Dale A. Mackall. "Flight Research Simulation Takes Off." *Aerospace America*, August 1993.

Schwarz, Frederic D. "1929 Flying Blind." American Heritage, August/September 2004.

Scott, William B. "FAA Approves Training by Simulation." *Aviation Week & Space Technology*, August 16, 1982.

Smith, Bruce A. "Flight Management System Adds Roles." *Aviation Week & Space Technology*, April 16, 1979.

Stein, Kenneth J. "Total Simulation Seen in Training." *Aviation Week & Space Technology*, September 8, 1980.

"Total Simulation Crew Training Pushed." *Aviation Week & Space Technology*, November 12, 1979.

Wallace, Lane E. *Two Decades with NASA Langley's 737 Flying Laboratory*. Washington, DC: NASA, 1994.

Chapter 6

Airbus Customer Services. *Flight Operations Briefing Notes: Wake Turbulence Awareness/Avoidance*. Toulouse, France: Airbus, June 2005.

Australian Transport Safety Bureau. "Mountain Wave Turbulence." Accessed August 13, 2015. https://www.atsb.gov.au/publications/2005/mountain_wave_turbulence.aspx.

Bibel, George. *Beyond the Black Box: The Forensics of Airplane Crashes*. Baltimore: Johns Hopkins University Press, 2008.

Brotak, Ed. "Bumpy Ride Ahead." *AeroSafety World*, March 2014.

Bullis, Kevin. "Folding Wings Will Make Boeing's Next Generation Airplane More Efficient." *MIT Technology Review*, November 18, 2013.

Chaney, Peter, Brent Jang, Dawn Walton, and Scott Norval. "15 Seconds of Terror." *Globe and Mail*, January 11, 2008.

Croft, John. "Weather Wars." *Aviation Week & Space Technology*, July 21, 2014.

———. "WSI's TAPS Weather System Could Provide Tracking." *Aviation Week & Space Technology*, June 8, 2015.

———. "Better Metrics Can Cut Turbulence Costs." *Aviation Week & Space Technology*, September 21, 2015.

de Lozar, Alberto. "An Experimental Study of the Decay of Turbulent Puffs in Pipe Flow." Philosophical Transactions of the Royal Society (February 2009): doi:10.1098/rsta.2008.0199.

Federal Aviation Administration. *Pilot's Handbook of Aeronautical Knowledge*. Washington, DC: Federal Aviation Administration, 2008.

———. "Why NextGen Matters." Accessed June 5, 2013. http://www.faa.gov/nextgen/why_nextgen_matters/.

———. *Aircraft Wake Turbulence*. Advisory Circular AC 90.23.6. Washington, DC: FAA, February 10, 2014.

———. *Aviation Weather Services*. AC 00-45G. Washington, DC: FAA, October 2014.

———. *Safety Briefing: Weather Forces, Sources, and Resources*. Washington, DC: FAA, March/April 2015.

———. "Turbulence: Staying Safe." Accessed August 12, 2015. https://www.faa.gov/passengers/fly_safe/turbulence/.

———. "Lessons Learned." Accessed September 21, 2015. http://lessonslearned.faa.gov/ll_main.cfm?TabID=1&LLID=66&LLTypeID=14

———. *Wake Turbulence Recategorization*. JO7110.65C. Washington, DC: FAA, February 29, 2016.

Feitag, William, and E. Terry Schulze. "Blending Winglets Improve Performance." *AERO Magazine*, Quarter 3.09, 2009.

Fong, Petti, and Emily Mathieu. "15 Seconds of Terror on Air Canada Flight 190." *Toronto Star*, January 11, 2008.

Hurt, H. H., Jr. *Aerodynamics for Naval Aviators*. Navair 00-80T-80. January 1965.

Job, MacArthur. *Air Disaster*. Vol. 4. Osceola, WI: Motor Books International, 2001.

Lelaine, Claude. *A380 Wake Vortex Working Process and Status*. Toulouse, France: Airbus, September 2008.

Moin, Parviz, and John Kim. "Tackling Turbulence with Supercomputers." *Scientific American*, January 1997.

National Center for Atmospheric Research UCAR Research Applications Laboratory. "Turbulence." Accessed August 18, 2015. https://nar.ucar.edu/2014/ral/turbulence.

National Oceanic and Atmospheric Administration. "NOAA Next-Generation Radar (NEXRAD) Products." Updated June 2, 2015. https://catalog.data.gov/dataset/noaa-next-generation-radar-nexrad-products.

———. "Terminal Doppler Weather Radar (TDWR)." Accessed August 18, 2015. https://www.ncdc.noaa.gov/data-access/radar-data/tdwr.

National Transportation Safety Board. *Aircraft Accident Report: Delta Airlines McDonnell Douglas DC-9-14, Greater Southwest International Airport, Fort Worth, Texas, May 30, 1972. NTSB-AAR-73-3.* Washington, DC: NTSB, March 13, 1973.

———. *Delta Air Lines, Inc., Lockheed L-1011-385-1, N726DA, Dallas/Fort Worth International Airport, Texas, August 2, 1985.* NTSB/AAR-86-05. Washington, DC: NTSB, August 14, 1986.

———. *Accident Report FTW93IA118.* Washington, DC: NTSB, December 3, 1993.

———. *Accident Report DCA98MA015.* Washington, DC: NTSB, May 14, 2001.

Potter, Everett. "Five Myths about Air Turbulence." *USA Today*, July 10, 2014.

Rosenkrans, Wayne. "Outmaneuvered Airflow." *AeroSafety World*, February 2013.

SKYBrary. "Airbus A380 Wake Vortex Guidance." Accessed June 5, 2013. http://www.skybrary.aero/index.php/Airbus_A380_Wake_Vortex_Guidance.

Thompson, Paul. "Total Turbulence Helps Guide Flights to Smooth Skies." *NYC Aviation*, April 8, 2015.

Thompson, R. G. "Dash 80." *Smithsonian Air & Space Magazine*, April 1987.

Triplett, William. "The Calculators of Calm." *Smithsonian Air & Space Magazine*, March 2005.

Vegano, Dan. "Turbulence Theory Gets a Bit Choppy." *USA Today*, September 10, 2006.

Warhaft, Z. *An Introduction to Thermal-Fluid Engineering: The Engine and the Atmosphere.* Cambridge: Cambridge University Press, 1997.

Chapter 7

Aggarwal, Arun K. *Computer Program for Analysis of High Speed, Single Row, Angular Contact, Spherical Roller Bearing, SASHBEAN Volume I: User's Guide.* NASA Contractor Report 191183. Washington, DC: National Aeronautics and Space Administration, September 1993.

Dornheim, Michael A. "A330 Fuel System: How It Works and Pilot Choices." Aviation Week & Space Technology, September 3, 2001.

ExxonMobil. *ExxonMobil Aviation Lubricants.* Irving, TX: ExxonMobil, 2009.

Fiorino, Frances. "A330 Overwater Flameout Raises ETOPS Issues." *Aviation Week & Space Technology*, August 31, 2001.

———. "Air Transat Fined for Maintenance Lapse." *Aviation Week & Space Technology*, September 10, 2001.

Ford, Kenneth W. "The Physics of Soaring." *Physics Teacher* 38, no. 1, January 2000.

Fraes, Paul Jackson, ed. *Jane's All the World's Aircraft 2007-2008*. Surrey, UK: Jane's Information Group, 2007.

Government of Portugal. *All Engines-Out Landing Due to Fuel Exhaustion*. Accident Investigation Final Report. Lajes, Azores: Government of Portugal, August 24, 2001.

National Transportation Safety Board. *Flight 30H, Boston-Logan International Airport, January 23, 1982*. NTSB-AAR-82-15. Washington, DC: NTSB, December 15, 1982.

Olsen, Michael, and Teddy Garland. "Object-Oriented Turbofan Engine Model (.NET)." Accessed October 22, 2012. http://echovoice.com/projects/engine.

"Report of Dual Engine Flameout on Trans-Atlantic Flight Raises Significant Safety Issues." *Aviation Today*, November 8, 2004.

Saarlas, Maido. Aircraft Performance. Hoboken, NJ: Wiley, 2007.

"Significant Regulatory Activity." *Air Safety Week*, January 9, 2006.

Chapter 8

Air Accidents Investigation Branch. *Report on the Accident to Boeing 777-236ER, G-YMMM, at London Heathrow Airport on 17 January 2008*. London: Air Accidents Investigation Branch, February 9, 2010.

Baxter, Martin, et al. "A Climatology of Snow-to-Liquid Ratio for the Contiguous United States." *Weather and Forecasting* (October 2005): doi:10.1175/WAF856.1.

Bibel, George. *Beyond the Black Box: The Forensics of Airplane Crashes*. Baltimore: Johns Hopkins University Press, 2008.

Boeing Company. *777 Flight Crew Training Manual*. Qatar Airways. April 28, 2008.

———. "Delta 777-200ER N862DA Thrust Rollback at Cruise: 26 November 08." Chicago: Boeing Company, February 28, 2011.

"Boeing Performs Crash Test on 787 Fuselage Section." Accessed October 8, 2015, http://komonews.com/archive/boeing-performs-crash-test-on-787-fuselage-section.

Hardman, Robert. "From My Window I Saw a Plane Coming In to Land. Suddenly It Exploded in a Cloud of Black Smoke, Flame and Debris HOW I SAW IT." *Daily Mail*, January 18, 2008.

Kaminski-Morrow, David. "Rolls-Royce: Trent Modification Will 'Eliminate' Fuel-Ice Risk." *Flight Global*, February 9, 2010.

Lockridge, J. "Energy Management: Important Even When Crashing." Aircraft Owners and Pilots Association. Accessed November 22, 2015, https://www.aopa.org/asf/publications/inst_reports2.cfm?article=5764.

National Transportation Safety Board. *Incident Report DCA09IA014*. Washington, DC: NTSB, January 27, 2011.

"Passengers Tell of Their Brush with Death." *Daily Mail*, January 18, 2008.

Williams, David. "Everyone Was Screaming. Kids Were Crying . . . We Thought We Were Going to Die." *Daily Mail*, January 18, 2008.

Chapter 9

Airbus. *A340-200/-300: Aircraft Characteristics. Airport and Maintenance Planning.* Toulouse, France: Airbus, October 2015.

Boeing Company. *Asiana Airlines (AAR) 777-200ER HL7742 Landing Accident at San Francisco—6 July 2013.* Chicago: Boeing Company, March 17, 2014.

Coker, Michael. "Why and When to Perform a Go-Around Maneuver." *AERO Magazine*, Quarter 2.14, 2014.

Croft, John, and Guy Norris. "Were Asiana Pilots Caught in the FLCH 'Trap'?" *Aviation Week & Space Technology*, July 22, 2013.

Huber, Mark. "How Things Work: Evacuation Slides." *Air & Space Magazine*, November 2007.

Jenkins, Marisa, et al. "Reducing Runway Landing Overruns." *AERO Magazine*, Quarter 3.12, 2012.

Milstein, Michael. "Double the Size of an Airbus A380?" *Air & Space Magazine*, July 2006.

National Transportation Safety Board. "Exhibit No. 2-1." In *Operational Factors*. Group Chairman's Factual Report DCA13MA120. Washington, DC: NTSB, November 15, 2013.

———. *Aircraft Performance Group Crash Site Factual Report.* Washington, DC: NTSB, December 6, 2013.

———. "Addendum 2." In *Human Performance*. Group Chairman's Factual Report DCA13MA120. Washington, DC: NTSB, 2014.

———. *Descent Below Visual Glidepath and Impact with Seawall Asiana Airlines Flight 214 Boeing 777 San Francisco, California. July 6, 2013.* NTSB/AAR-14/01. Washington, DC: NTSB, June 24, 2014.

Pasztor, Andy, and Rolfe Winkler. "SpaceX Puts Satellites into Orbit in Its First Launch Since Explosion." *Wall Street Journal*, January 14, 2017.

Rosenkrans, Wayne. "Over in a Flash." *AeroSafety World*, January 2007. https://flightsafety.org/wp-content/uploads/2016/12/asw_jan07_p46-49.pdf.

Taylor, Rob, and Rachel Pannett. "Three-Year Search for Malaysia Airlines Flight 370 Ends Where It Started: Shrouded in Mystery." *Wall Street Journal*, January 17, 2017.

Transportation Safety Board of Canada. *Landing Overrun and Fire 02 August 2005.* Aviation Investigation Report A05H0002. Gatineau: Transportation Safety Board of Canada, October 16, 2007.

VartAbedian, Ralph, and Denise Gellene. "A Dangerous Test at McDonnell." *Los Angeles Times*, November 6, 1991.

Epilogue

Airbus. *Commercial Aviation Accidents 1958–2014: A Statistical Analysis*. Toulouse, France: Airbus, May 2015.

Boeing Company. *2015 Statistical Summary of Commercial Jet Accidents Worldwide Operations 1959–2104*. Chicago: Boeing Company, July 2016.

International Civil Aviation Organisation. *ICAO Safety Report*, 2015 ed. Montreal: International Civil Aviation Organisation, 2015.

Index

229–30, 233; stall, 234; wake turbulence, 186, 193; wing testing, 114. *See also* Asiana Airlines Flight 214; British Airways Flight 38; Malaysia Airlines Flight 370

Boeing 787, 19, 26, 47, 48, 58, 137, 153, 179, 234, 270, 280

Boeing KC-135 Stratotanker, 129

boundary layer, 147, 148

brakes, 54, 58–61; antilock, 65; auto-, 64, 65, 80, 277; carbon, 59–60, 75; certification tests, 73, 74–75; fade, 58, 59; failure, 72–74; fire, 65–67, 74, 227; life, 74; temperature sensors, 70, 75; wear, 73–74

British Aerospace BAe–146, 101

British Airways Flight 38: cavitation, 242, 246; data mining, 245; flaps, 229–30, 233; fuel system, 242–45; ice, 240, 241, 243, 245, 246, 250, 251; landing gear, 234, 236–38; sinkrate, 232; stall, 234

buffet, 104, 139, 140, 146, 149–50, 157, 233

calibrated airspeed (CAS), 151

cavitation, 242–43, 246

center of gravity, 24–29, 30, 38–39, 40, 51, 74, 98, 105, 106, 203

certification testing: brakes, 70, 74, 75, 273; engines, 50; evacuation, 275–76; fuel system, 245; takeoff, 77; wings, 114, 200

CF6-50E2, 21

CF6-80C2, 285

checklist discipline, 222

clear air turbulence, 201, 202, 203, 204, 208

cockpit voice recorder (CVR), 3, 41, 69, 70, 86, 148, 159

computer simulation, 88, 115, 161

Concorde, 14, 69, 153, 285

control law, 154–56

control yoke, 83, 98, 103, 104, 152, 154, 157, 169, 234, 250, 287

coordinated turn, 88–89, 108–18, 121–22, 123

crew resource management (CRM), 34, 38, 88, 96, 97, 169

cross-check, 87, 89, 91, 95

CVR. *See* cockpit voice recorder

direct lift control (DLC), 102

drag, 14, 54, 79, 100, 101, 107, 116, 120, 150, 167, 170, 203, 217, 229, 231, 249, 267; equation, 22; ground effect, 192; high-speed buffet, 104, 140, 149–50; induced, 192; landing gear, 253; lift-to-drag ratio, 217–19, 226, 259

drop test, 234

Dutch roll, 122, 126–30

elevator, 28, 29, 39, 82–83, 98, 99, 100, 103, 104, 105, 150, 209

energy: air, 151; approach, 227; brake, 58, 60, 74–75, 227; compressor, 44; crushable cartridge, 21; exhaust gas, 47, 48; explosive decompression, 143; kinetic, 236, 238; landing, 270, 276; potential, 250; rejected takeoff, 68–69, 74; rolling resistance, 56; rubber, 60; shockwave, 126; tire, 67; turbine, 45

ETOPS (Extended Operations), 215, 288–89

FAA. *See* Federal Aviation Administration

FANS. *See* Future Air Navigation System

fatigue. *See* metal fatigue; pilot fatigue

FBW. *See* fly-by-wire

Federal Aviation Administration (FAA): Dutch roll training, 130; Kalitta Air, 43; Kenya Airways Flight 507, 86; NextGen, 134; simulator, 161, 180, 181; *Takeoff Safety Training Aid*, 75; transponder, 132

flaps, 51, 99, 100, 107, 124, 167; Airbus A320, 55; Airbus A340, 271; asymmetric, 72; Boeing 747, 10–12, 13; Boeing 777, 229–30, 231, 234, 248, 249, 250, 252, 255; Lockheed L-1011, 102; rejected takeoff, 64

flight control computer, 154, 170

flight data recorder, 3, 50, 78, 86, 239

flight management system (FMS), 177

flight mode annunciator (FMA), 95, 174, 175, 266, 270

flow separation, 146, 147, 148, 150, 233